SOTHIS PRESS

Sothis Press aims to completely reimagine the public domain book with an emphasis on universal access to captivatingly designed print editions. We believe that technology will enable every single book ever published to be perpetually accessible in beautiful print form. Sothis is beauty in ubiquity.

www.sothis.press
1 Waterhouse Square, London, ECIN 2ST

The Principles of Automobile Body Design

Covering the Fundamentals of Open and Closed Passenger Body Design, With Chapters on the Design of Commercial Bodies

Kingston Forbes

SOTHIS PRESS

ISBN 978-1-0368-0372-8

The Principles of Automobile Body Design is in the public domain.

This edition is published by Sothis Press © 2023. The company is registered in England and Wales. The Company Number is 14900389.

The Principles of Automobile Body Design

Covering the Fundamentals of Open and Closed Passenger Body Design, with Chapters on the Design of Commercial Bodies

By KINGSTON FORBES, M. E.

Illustrated with six hundred twenty-three half-tone plates and line drawings

1922
WARE BROS. COMPANY
1010 Arch Street
PHILADELPHIA, PA.

CONTENTS

THE PRINCIPLES OF AUTOMOBILE BODY DESIGN

The lack of information on body engineering and the demand for reprints of the articles which were published in "Motor Vehicle Monthly" led me to re-arrange and re-write these articles so that they could be incorporated in book form. Body engineering has made tremendous strides in the last few years and the writer has endeavored to collect all the data possible which would interest the body engineer and the student.

The possibilities of this profession are very good, and it is hoped that the book will be of help to the student as well as the engineer. The scope and aims of the body engineer were outlined in a paper by the present writer, read before the Society of Automotive Engineers, in New York, January 12, 1921, and the following extracts will make this preface complete:

THE BODY ENGINEER AND HIS RELATION TO THE AUTOMOTIVE INDUSTRY

In choosing the title for this paper it was hoped that a note could be struck which would bring to the attention of the industry the broadness and scope of body engineering. Also to outline the way this side of the industry can best be considered and developed. Body engineering has, of course, to look for mass production or big business for its greatest encouragement. This is the age of big business, and the automobile industry *is* a big business, in fact one of the biggest, and not far from being actually the biggest in the country. Big business makes demands for organizations which the small business does not need. Small items in a small scale production can be easily handled, or they about handle themselves. Small items in great production are of tremendous importance, for as a chain is as strong as its weakest link, so is big business.

If a thousand cars are made in a day and a shortage occurs in one of the smallest items it is possible that the entire production is held up, incurring the loss of thousands of dollars. This emphasizes the point of importance of every detail in large business in a way which no one can fail to grasp. For instance, if a change in design could be made by the body engineer that would permit cutting the leather for the trimming to 15 per cent. waste instead of 20 per cent., a saving of $1,000 and over per day would be made on a car production of 1,000 cars per day. In a small shop or custom shop a few feet of leather more or less would not make a difference, but it means so much in a large production that this and every other detail must be considered very carefully.

Body engineering is really a broader field than the title conveys, as will be shown in the outline of this paper. The body engineer's relation to a body manufacturing plant, a large car manufacturing plant, a small car plant and a custom shop demands different classes of engineers. In the main, the body engineer is responsible for the external appearance of the entire car. If it is a custom-built car, a preliminary sketch of the complete car is made and very often this is all the body builder, top maker and trimmer has to work on, so he gets the car to look as nearly like the sketch as possible. A body plant very often submits designs for a complete car, but generally has only the body to build. In order to work out his ideas or combinations of ideas the body engineer of the small plant relies, to a great extent, on the companies which make his bodies. In a large plant where bodies and all other metal parts are made, the body engineer has to consider the manufacturing details involved in all parts that he designs.

Just to cite an instance: There is a distinct difference between the mechanical side of the automobile and the general appearance or the artistic side. No one would ordinarily combine the story writer and the illustrator together in more than a co-operative spirit. The success of the story does not depend upon the illustrations, or *vice versa*. But having a good story properly illustrated gives a wider field for its sale.

So it is with a car; the mechanical condition and limitations have to be taken in hand by the body engineer, and it does not stretch the point at all in saying that its first qualification must be an artistic one. The proof of this statement is demonstrated in the 1921 cars now being displayed. The body engineer, for the company specializing in custom body work, must apply artistic principles to every car he has to create, each car, calling for individuality of its own, draws heavily on his power of creation. But the body engineer for the large production has to satisfy 100,000 people with perhaps one design. There is a distinct difference between that and satisfying the individual as the custom engineer must. Now we know that the tastes of 100,000 people are not all the same even though some of our movie stars have practically an unlimited following. The design then for the 100,000 is not the bizarre design, but one that will come as near as possible to being acceptable to the majority.

It is often remarked that the so-and-so car does not quite suit the person talking and invariably he remarks that there is a certain characteristic that he favors. Take several such individuals and see what all their ideas pertain to and it will be found none will agree on any specific design. But if a car of established popularity is shown practically all will be satisfied with it. The point to make is that the simplicity of the car's design and absence of jarring notes can only be evolved by painstaking effort. The layman can not appreciate the work behind the smooth finished product, as the effort is not as apparent as the finished carvings on historic buildings and churches that show the effort required by their intricately embellished carvings.

Art in the generally accepted term applies to pictures, carvings, etc. Art as applied to an automobile has to be manifested in the arrangement and shaping of sheet metal units on frames. The requirements of art have one other standard to match up to, and that is, can the result required be accomplished by economical manufacturing? One thousand cars or more a day do not allow for fancy hand decorations or carvings. Art has to be satisfied by huge metal presses and metal-forming machines. To achieve results the practical side of the body engineer must be as heavy as his artistic side, as they must work hand in hand to avoid disappointment. While we have definite lines upon which the artist bases his work, his success does not depend upon geometrically arranged lines, nor can the composition of his pictures be worked out with a slide rule. The mechanical side of the car is quite amenable to slide rule practice, but the body lines are not. This is one of the reasons why body engineering has an indefinite position in the industry.

In large production it is not even possible to control the metal parts to a very fine degree owing to variations in the grades of metal and fluctuations in the presses. The upholstering, top design and painting, including enameling, all have bearings on the final results and have to receive the attention of the body engineer. The main trades which govern the body engineer's work are wood working, sheet metal working, metal machine work, braces and hardware, trimming, top building, painting and enameling. As the practical side of the engineer is emphasized, he should be familiar with all these trades and they are all peculiar in being professions that can not be learned by correspondence courses. In fact there are practically no schools or instruction books covering the modern phases of this work.

The body engineering department could be divided into six main divisions as shown in the following layout:

1. Body construction, open and closed.
2. Sheet metal, body metal, fenders, hood, radiators, etc.
3. Trimming.
4. Top building.
5. General hardware and mountings.
6. Painting and enameling.

All these arts covered by the six divisions dovetail into one another and must be considered in the preliminary design. For instance, the radiator and hood can really be considered the starting points for the car body design, and when these are established the main body construction is worked out for the dash or cowl line back.

The exterior lines of the body must be pleasing to the eye, at the same time the passenger capacity has to be arranged to give ample room and comfort to the required number of passengers that the body is being constructed for.

In order to give ample comfort, satisfactory trimming has to be arranged for. Comfort depends on leg room, body room and positions, and well arranged cushions and back springs with soft padding, with suitable cloth, plush and leather covering.

Passing on to the top, if it is an open body the lines and windshield must be considered in order to permit a top to be made that will give adequate protection and at the same time be pleasing in appearance and of course harmonize with the rest of the car.

The fenders and side aprons and radiators and mud pans can add or detract a lot from the car's appearance, and must be worked out to blend with the other larger elements in the car craftsman's shift.

Under general hardware comes a lot of small parts, such as window regulators, door locks, windshields, hinges, etc., not very large in themselves but important in providing ease of operation and satisfactory service to the car owner. What is there more annoying on a car than defective locks, rattling doors, and hard-working or leaky windshields. And last but not least comes the painting, and when this is properly done it adds tremendously to the final results. In working out the design the panels, molding and corners should be made so it will simplify painting. Modern mass production methods use sprays and flowing operations, and if there are bad corners, mouldings and holes, these will seriously interfere with the paint, and cause varnish and dirt runs. If it is planned to have molding arranged in different colors from the main panels they must be arranged to give harmony and not produce discordant lines.

In the foregoing I have endeavored to show the broadness of the body engineering field of effort, and as the production increases the detailed accuracy required along these lines becomes more and more necessary. It will require organized departments to be able to handle this work. The grouping previously referred to can be made into department divisions. This can be controlled by a separate engineer, if the amount of work warrants it, and one or more men to keep the work in progress. A general drafting force with chief draftsman and checker are also required.

Modern accounting and production system demands detailed information about every piece which goes to make up a car, and this entails a lot of detail drafting work. Every piece of wood, top material, side curtain material, reinforcement and wadding, trim material, leather, imitation leather, binding; ounces of hair, tacks, nails, etc.; every screw, nail, pieces of hardware. This means engineering information is required with blue prints and specifications. The old method used to be to make paper patterns of all parts and let the manufacturing departments control their own work, and whenever duplication of parts was required, additional paper patterns were made and sent out. Long ago it was found that no progress could be made in manufacturing without proper engineering records and blue prints.

I venture to say that there are several modern auto plants that have no definite engineering control of the material purchased and fabricated in the groups outlined, and this material runs into millions of dollars per year.

While the art of coach building is as old as the ages, it was brought to a manufacturing basis only a few years ago when buggy production hit its high mark. The advent of the automobile brought up entirely new problems and the automobile body engineer is practically a recent accession to the engineering profession. The importance of the body engineer can be measured by his designs and by his control of the materials and fabrication of the material required on the automobile body. It is not sufficient that a beautiful and readily marketable design be produced, for it must also be possible to make it economically and without waste of material.

The future holds out big opportunities for the body builder and engineer. Competition will be keener than ever, not only among the body manufacturers, but also among the car manufacturers. The automobile competition will be the body engineer's opportunity. With the stabilizing of chassis construction, the external body and allied construction will become one of the biggest factors in marketability and stimulation of sales.

Painting comes under body engineering, but the subject is such a broad one that it would require a volume by itself, so it was thought advisable to omit it from this book.

KINGSTON FORBES.

March, 1922.

THE PRINCIPLES OF AUTOMOBILE BODY DESIGN

CHAPTER I

EVOLUTION OF THE AUTOMOBILE BODY

Since the time when the first gasoline engine was adapted to the road vehicle, the progress of development of the mechanical features of the automobile has been rapid and sure, while the exterior features of the car, which can be classed as the running gear and the body, did not keep up with the development of the chassis.

When the chassis and engine had been brought up to a high degree of efficiency, attention was brought to bear upon the body and other features of the car that had any reference to the comfort of riders and the improvement of the car's external appearance.

The possibilities for many improvements were soon noted and acted upon, the result being that the automobile departed from the ways and appearances of the horse-drawn vehicle and took on a distinctive style of its own.

The change was revolutionary, as the passenger-carrying conveniences had been designed by basing the construction on the experience gained in making horse-drawn vehicles, and the new design and construction called for altogether different conditions.

A new industry of vast proportions was created which had to develop new machinery and new methods to cope with the requirements of the automobile body and its component parts. The methods used to manufacture only a few years ago seem crude when compared to the methods that are now in vogue; also, the costs were high compared to the new costs. New industries have sprung up so quickly around some of the component parts that the great majority do not know about them, as few have had time to write about or describe them.

The various phases of development in the body and the running gears, as the other parts are called, will be dealt with briefly in this volume. These parts have now arrived at the stage where improvements are hard to make and new designs are hard to create, and when this stage is reached, analytical study is required to accomplish any further improvements. Progress is generally made by basing new developments on previous practice, and if the early developments are not known, the work is carried on upon an insecure foundation.

The aim of this book is to furnish the foundation so that past results can be considered while new ideas are being developed. That the time is here when a

FIG. 1.
A typical body of about the year 1903

new field of effort is opened up is demonstrated by the cars of 1921 and 1922, as shown at the automobile shows. These shows could be classed as body shows, because the important developments shown in the cars over the previous season

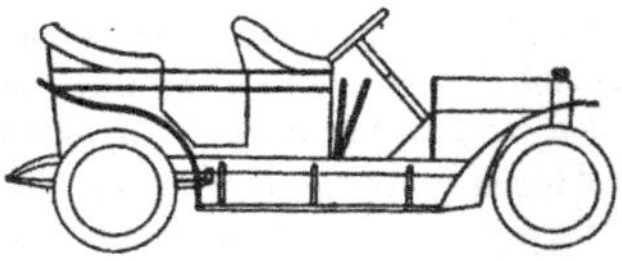

FIG. 2
A typical body of about the year 1905

were in the refinement and improvement of the body design, running gear and equipment.

America's methods of manufacture have done a great deal toward the development of the automobile, as they have produced cars in vast numbers and at a

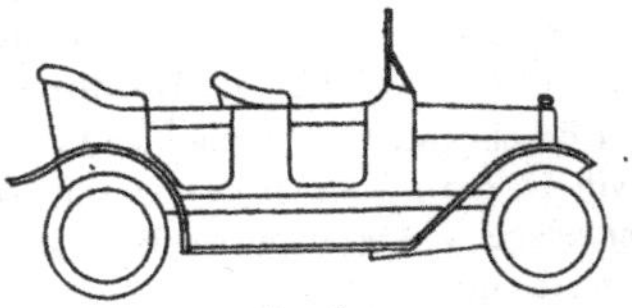

FIG. 3
A typical body of about the year 1908

price that brings them within the sphere of the majority of people. The result is that the cars are built in great quantities of one design, as this makes it possible to develop ways and means to produce cheaply. The development of the automobile must necessarily hinge around that class of car that manufactures in quantities, and not around the special hand-made production.

It requires a great deal more work to develop a car for quantity production than it does to make one special car. Every detail must be considered as every penny counts when the quantity produced runs into the thousands. There are greater possibilities in the design and production of the manufactured car, consequently most of the work in this book has been done with large production in mind,

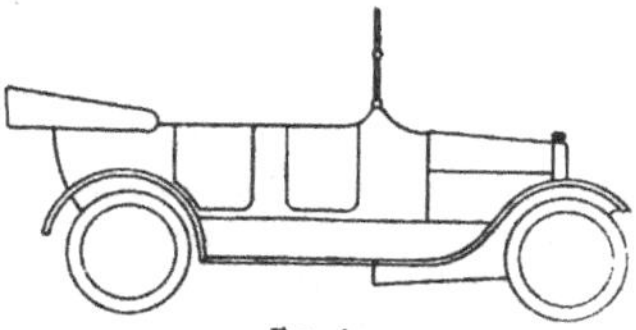

Fig. 4
A typical body of about the year 1912

but of course there are many of the problems which apply just as well to the sample or custom car as they do to the quantity production one.

It is very interesting to trace the development of the automobile body from its early stages up to the present beautiful streamline designs. The first model, dated about 1903, consisted of a box-like structure with two seats on it, conveni-

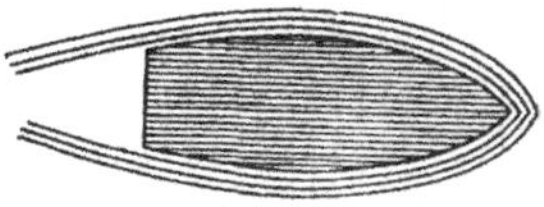

Fig. 5
"Boatline" shape

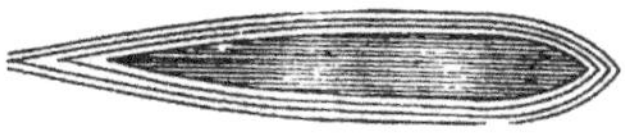

Fig. 6
"Streamline" shape

ently located adjacent to the steering lever or wheel. In Fig. 1 is shown a line sketch of a typical body of this period. This would be considered in present-day terms as a roadster, and is the prototype for our present two-passenger cars. Between the above period and 1905 a variety of body types were made and accom-

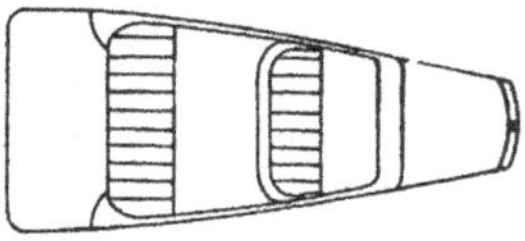

Fig. 7
Representative shape of a touring body

modations for two additional passengers provided. Some of these four-passenger bodies had a door at the rear for extra passengers to get in and out; this did not prove satisfactory, as it cut down the room for a back on the rear seats, and the passengers had to sit face to face, making it almost a jaunting car effect. In Fig. 2 is shown a sketch of a typical car of this period. It will be noted that a great many corners and mouldings are provided. The design savors a good deal of the horse-drawn vehicle, to which a buggy type of top is fitted. The car shown for the 1908 period in Fig. 3 has a body that is very similar to the 1905 period; the only difference is the slight improvement in outline and detail, and a smoother line is shown throughout the design. In the 1912 style shown in Fig. 4 is seen the advent of the "four-door" type, and a windshield fitted as permanent equipment. The four-door type of body marks the coming of an "automobile" body and the breaking away from the buggy and kindred horse-drawn vehicle types. This body has capacity for five passengers, and has four doors.

The 1914 models show the streamline type of body. This has five-passenger capacity and four doors that operate. The windshield brackets are set on the body and make the windshield a permanent part of the car. The dash is made smaller and the hood larger, so that the two are of the same size and present an unbroken line from the radiator to the back of the body. The name "streamline" dates from this period when the bodies were made to match the front of the hood and present

FIG. 8
Approximate type of modern body

a clean, smooth appearance. The title "streamline" has been much abused, as any body that eliminated the offset hood and dash took that name. Boatline would be a name that would be more descriptive of the style, as the front end of the car is always the narrowest part, as in a boat shape. See Fig. 5, which illustrates a boat shape. A true streamline shape would take on the proportion of a fish or torpedo, the widest parts of this being the front end, as shown in Fig. 6. The theory of a true streamline body is that it will go through the atmosphere with the minimum degree of resistance; the air displaced by the maximum cross sectional area must be permitted to close back again without creating a vacuum. This is accomplished by the tail-like end of the fish or torpedo body. This is illustrated by the lines that are shown outside of the torpedo-like shape in Fig. 6. These lines represent air, and show how the air is parted easily and allowed to flow back to the natural condition gradually. In Fig. 5 the lines will be seen to leave a gap at the end of the body. This is a vacuum that this style of body would create, and has a tendency to pull the body back. In Fig. 7 is shown a representative shape of a touring car body. This, it will be noted, has a decided wedge shape, and a vacuum-creating back. It is obviously impractical to adapt the torpedo-like shape to a touring car body, so that as a comparative name the term streamline is a misnomer.

For the 1916-17 period, Fig. 8 shows the approximate type of car that is so familiar now. This has the long, straight lines, with a hooded front set back.

This tends to smooth down the necessary projection of the seat back and give the appearance of a continuous line along the top of the body. The projecting windshield brackets are also eliminated, so that the windshield attaches to the body without any unsightly projections. This brief review of the body developments describes the periods when these changes went into general effect on the American cars. Some makers can claim that they had any one of these types several years previous to the times stated, and this may be true in a few cases, because up to a few years ago the European designs were followed by the majority of manufacturers. Consequently the general American styles at the beginning of the great war were several years behind the European ones. Since that time we have more than caught up with European design.

Fig. 19 Pierce-Arrow

Fig. 20 Packard Twin-Six

Fig. 21 Cadillac

Fig. 22 Buick 22-45

CHAPTER II

MODERN TYPES OF OPEN BODIES

The modern type of body is, without question, a so-called streamline one, and these lines are long and sweeping, with a clean cut line without any projecting portions or surfaces. A body can conform to this description and yet be made in a variety of forms. The most modern body is a degree of design which has no rules by which its merits can be differentiated. The variations from the long, sweeping lines are many, and it is individual taste that determines their preference.

In studying the body types that are marketed by the manufacturers for the

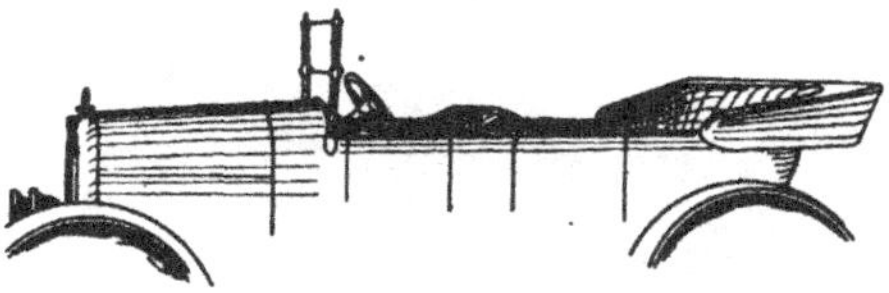

Fig. 9
The full-rounded type

last two or three seasons, it is found possible to divide the types into five divisions, viz: The first, the full-rounded type; second, the semi-rounded type; third, the angular type; fourth, the semi-angular, and in the fifth are grouped the miscellaneous types that have retained characteristic lines of their own.

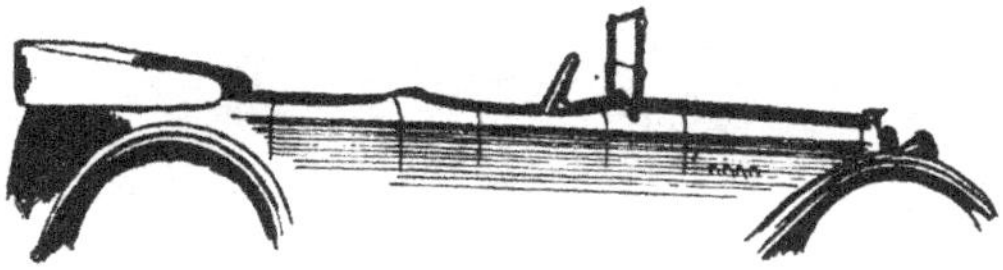

Fig. 10
The full-rounded type

In analyzing these groupings it is found that the full-rounded and semi-rounded types comprise 70 per cent. of all models, while those that retain characteristics of their own, in class five, comprise about 20 per cent. The other 10 per cent. are angular and semi-angular. If numbers proved the popular type, the full-rounded type would be called the most popular, but we find that in the small 10 per cent. class the designs are all new ones, which would probably indicate the trend of future design.

Fig. 23 Franklin Sedan

Fig. 24 Lafayette

Fig. 25 Lincoln

Fig. 26 Wills-St. Claire

To illustrate the types outlined, two representative cars in each group will be shown in Figs. 9 to 18. Taking up the first group, termed the full-rounded type, Figs. 9 and 10 show typical examples in this group. Starting from the radiator, which has a round front and a rounded top, the curves are carried along the hood,

FIG. 11
The semi-rounded type

shroud and the sides of the body, almost to the back of the seats. This type predominates, and 40 per cent. of the cars use these lines.

In the semi-rounded type, the radiator is round in front, and generally the top of the hood and the shroud carry out the same round lines. But the decided

FIG. 12
The semi-rounded type

curves stop at the front door, the sides of the body having a flat edge, and in some cases a bead or a small rounded corner. This type accounts for 30 per cent. In Figs. 11 and 12 are shown typical designs in this class.

FIG. 13
One of the "new" angular designs

The angular type of designs which are shown in Figs. 13 and 14 are "new" designs, the predominating feature being a decided tendency to flat surfaces and sharp corners. In Fig. 13 it will be seen that the radiator has sharp corners at the sides and at the top, and that these sharp corners are carried all the way back along the body. In Fig. 14 the sharp line starts at the radiator and extends as far as the windshield, and then runs up; the back of the body has angular corners

FIG. 14
The sharp line starts at the radiator

as well. In the semi-angular group, Figs. 15 and 16 are presented, and show the predominating body lines to be long and straight, with a small rounded corner. These have proved very popular designs.

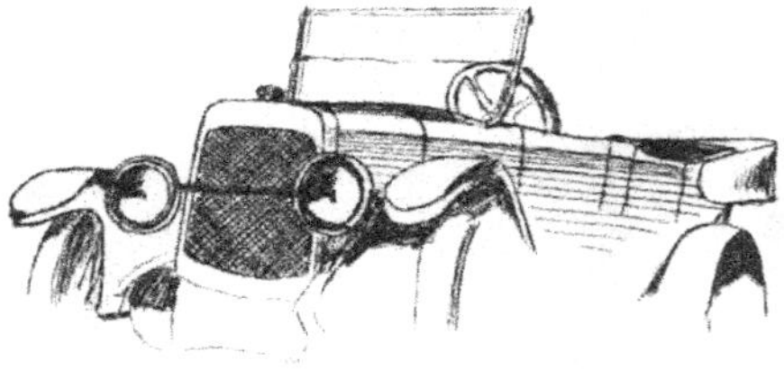

FIG. 15
This belongs to the semi-angular group

Under group five are shown Figs. 17 and 18, representing the Pierce-Arrow and Packard, but since these sketches were first prepared both of these companies have changed their body lines considerably. In making their changes they have

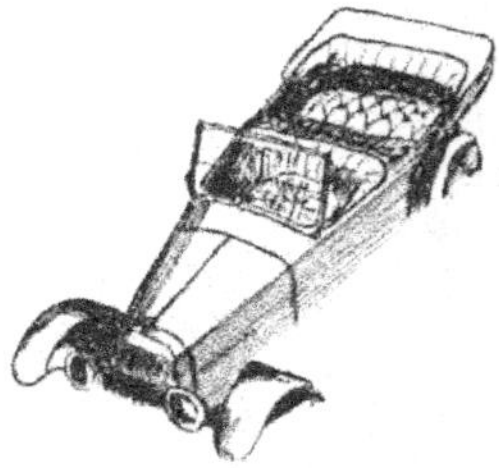

FIG. 16
The predominating body lines are long and straight

retained one of the most prominent features of their individuality, and that is the radiator shape. In both cases the radiator shape stamps them as well as their names, so familiar with the design is the public. As soon as you leave the radia-

Fig. 29 - Overland

Fig. 27 - Ford

Fig. 30 Buick 22-35

Fig. 28 - Dodge

Fig. 31 - Allen

tor and hood the body takes on the lines of the semi-angular groups. From a body standpoint it would be difficult to find many cars today with enough peculiarities to really class them in the miscellaneous group. However, it will serve a useful purpose if we put the ledge type bodies in this fifth group, starting with the Packard, Willys-Knight, Paige, and Essex. If we were analyzing foreign cars there would be a large field to select from for inclusion in this group.

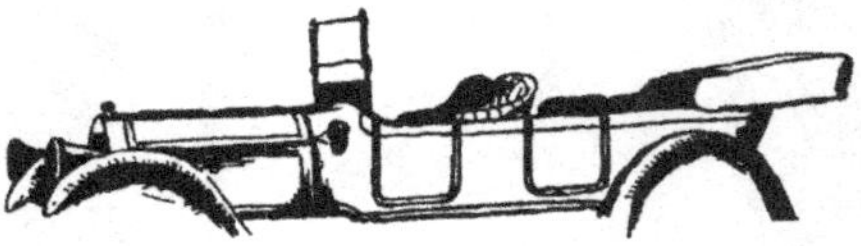

FIG. 17
Pierce-Arrow body lines of a few years ago

"The body designs of 1922" would perhaps be the best way to designate the photographs from 19 to 31. While they are modern in 1922, when this book is being written, in a few years they may be out of style. So the pictures may serve as a record of the types made in 1922. Figs. 19 and 20 show the latest designs of Pierce-Arrow and Packard. It will be noted that they have similar characteristics to the old design shown in Figs. 17 and 18. The Cadillac shown in 21 typifies a substantial car, well balanced, without any radical features whatever. In Fig. 22 is shown a Buick, a popular car in the medium priced field. It is noteworthy by its appearance of grace combined with substantiability. Fig. 23 shows the Franklin sedan, which is an air-cooled engine car, so it has slightly different radiator and hood lines, which tends to give the car a different appearance from the average car.

FIG. 18
Packard body lines, about 1916 period

Figs. 24, 25 and 26 show the Lafayette, Lincoln and Willis-St. Claire cars, all of which were developed and put on the market in 1920-21. It will be noted that they are all conservative in their body lines. All the cars shown have been 6, 8, or 12-cylindered cars of a good wheelbase. It is harder to get a good appearance with a short wheelbase four-cylindered car, so to make this fact more clear, photographs of the Ford, Dodge, Overland, Buick and Allen are shown in Figs. 27 to 31. These are popular cars and show well what lines can be worked out with short wheelbase and with a conservative cost in mind. If a long wheelbase is used, as it is on some high-priced cars of four cylinders, the body problem becomes identical to the 6, 8, or 12-cylinder car.

CHAPTER III

THE PRINCIPLES OF BODY DESIGN

It has been shown in the preceding chapters how the designs of bodies for the automobile developed up to the present pleasing type, and in the succeeding chapters are shown how the present general designs vary to conform with the circumstances governing the limitations of the design and capacity of the body.

The propelling and carrying structure of the automobile, which comprises the engine, frame, spring, axles, etc., are assembled into a group which is termed the chassis. The chassis, then, is that part of the vehicle that the body must be adapted to.

The body of the car starts at the back of the engine, and rests on the frame from the dash backward. The design is obviously governed by the chassis. Knowing the operating ability that is required, the engine and chassis proportions, the tire size and wheelbase are determined to give the desired ability. When this is done, the body, hood and fenders are designed to give the best appearance possible,

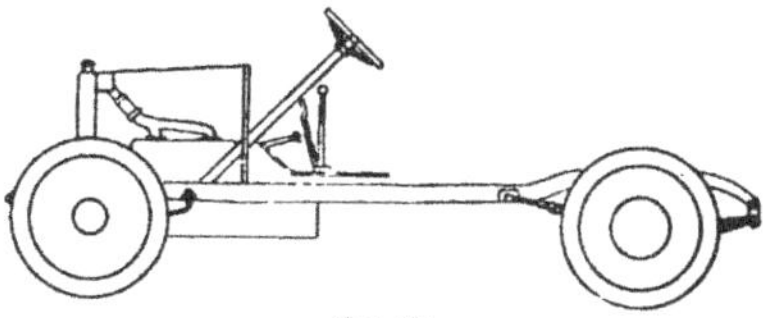

FIG. 32
A characteristic chassis layout

at the same time to give the maximum of possible comfort. The proportioning of the body becomes the principal consideration, as its design governs the proportions of the other parts that go to make the complete car appearance.

In Fig. 32 is shown a characteristic chassis layout, and it is apparent that the body must be back of the engine and that it may extend to a convenient distance past the rear wheel. The front seat is located to provide a comfortable seat in relation to the steering wheel and the hand control levers. On a touring car design the available distance at the rear of the front seat is to be used to provide a comfortable rear seat. Comfortable seats can be made in a variety of ways, and the room available will govern the methods used to obtain them. In Figs. 33, 34, 35 and 36 are shown average seating conditions under different circumstances. It will be noted that the higher the seats are, the shorter the distance is from the dash to the seat back, and the lower the seat is, the longer this distance has to be.

The average seating position for a five-passenger car, taken from a large number of cars, would be the condition shown in Fig. 37, and a seven-passenger seat-

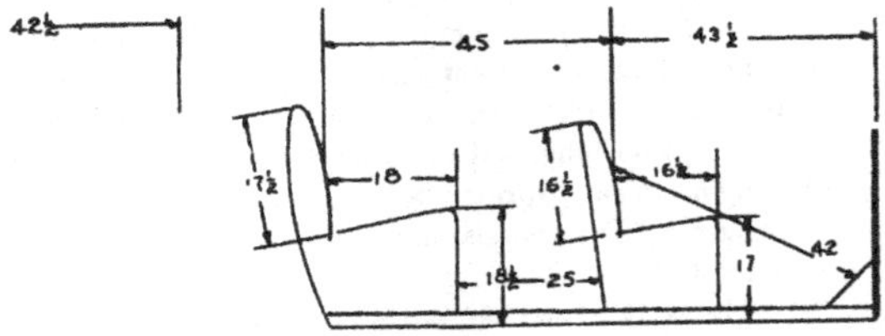

FIG. 33 FIG. 34

Figs. 33 to 36 show average seating conditions under different circumstances

ing position as shown in Fig. 38. The seating conditions have to be worked out to suit the cars, and these vary a great deal because of the different types and sizes of engines used.

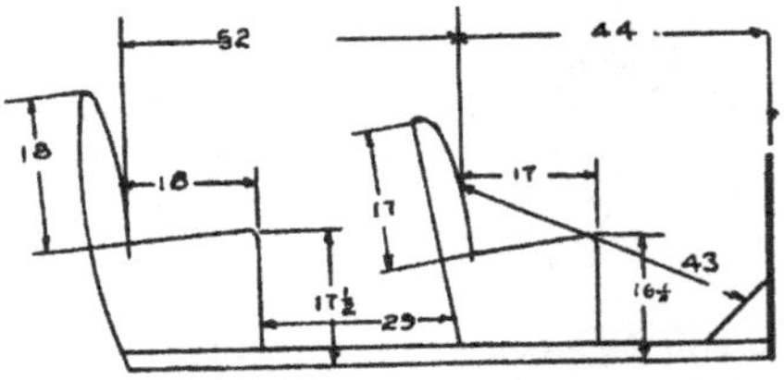

FIG. 35

There are four main types of engines in use, the four, eight, six and twelve, and these all give different conditions that affect the body design. In order to properly analyze these conditions the diagram in Fig. 39 has been made, and this

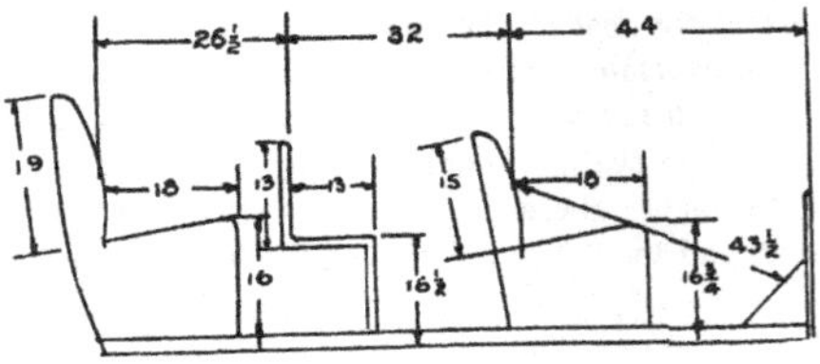

FIG. 36

shows an outline of a typical four-cylinder car. The main conditions that affect the body design are the parts that are divided off at A, B, C and D. These symbols are used so that a number of different cars can be measured and tabulated under these letters. The figures for three popular four-cylinder cars and two eight-cylinder cars are shown under this form:

No. of cyl.	A	B	C	D
4	79	18½	27¼	106
4	76	20	28	104
8	91	26	31	122
8	89½	24	30½	120
4	86½	20	28½	115

No. of cyl.	A	B	C	D
6	88	28	37½	125½
6	84¾	26½	37¼	122
6	87¼	26	36¾	124
12	87	26¾	35	122
12	95	26¾	40	135

In Fig. 40 is shown a typical outline for a six-cylinder car. The outline of the hood is noticeably longer, and the wheelbase increases accordingly. Three different six-cylinder cars and two twelve-cylinder cars have been measured up and the dimensions indicated.

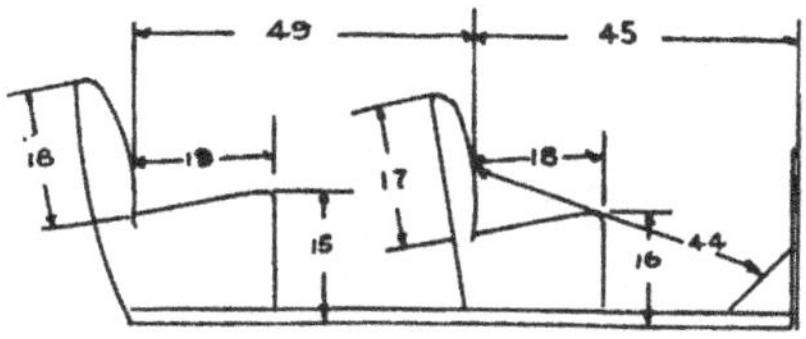

FIG. 37
Average seating position for a five-passenger car

By comparing the figures of the four-cylinder cars against the eight, it will be noted that the lower figure being 76, and the higher 91, there is a difference of 15 inches in dimension A. In dimension B the difference is 7½ inches; in C

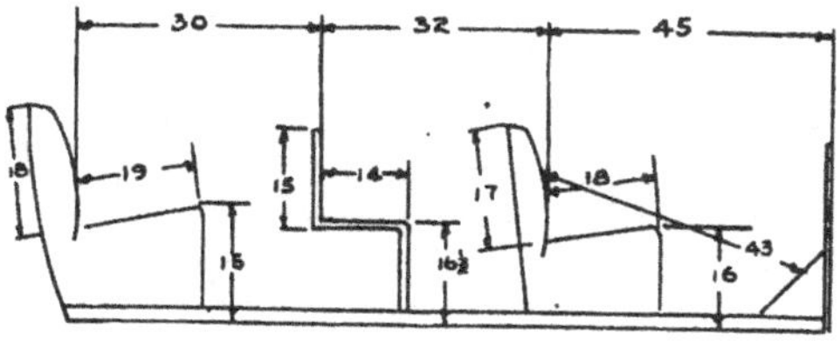

FIG. 38
Average seating position for a seven-passenger car

there is a difference of 3¾ inches, and in the wheelbase the difference is 18 inches.

In the same way, by comparing the six-cylinder car against the twelve, we note a difference of 10¼ inches, the lowest being 83¾ inches and the highest 95 inches. Dimension B varies 2 inches, and dimension C, 5 inches; the wheelbase, 13 inches. The main body dimension A runs from 76 inches on a four to 95 inches on a twelve, this big difference being caused by the four-cylinder car being kept as small as possible, while the twelve is made as large as possible, which accounts for the great difference of 19 inches. A good average dimension for A would be 86 inches, and would apply on a moderate wheelbase on either four or six-cylinder cars.

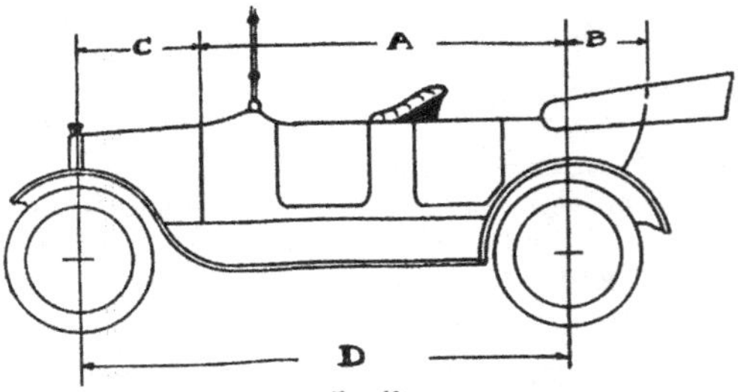

Fig. 39
Outline of a typical four-cylinder car

The seat and leg room having been considered, the width of the body that can be obtained will be defined. The front seats on touring cars and roadsters are generally constructed to accommodate two passengers, and have a width of 40 inches. The modern body lines tend to make the front seat wider, and so the seat width may run more around 42 inches. In the roadster the front seats are made even wider, so that three people can get in it when necessary. The rear seat on a touring car is generally made to take three people, and to do this comfortably as much space as possible is provided in it. The available space is limited by the distance between the rear wheels.

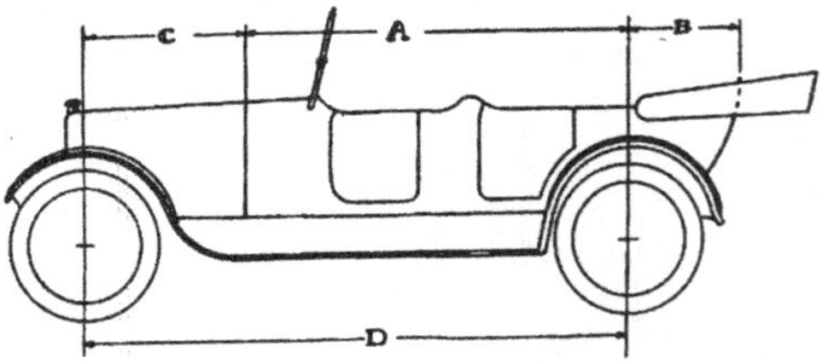

Fig. 40
Typical outline of a six-cylinder car

The standard track width for American cars is 56 inches, that is, the distance between the centers of the wheels. In the diagram, Fig. 41, the rear end of the body is shown, and also its relation between the rear wheels. It is seen that depressions have to be made in the sides of the body to clear the wheels. These depressions are called " wheelhouses," and have to be made about 48 inches wide. If the wheelhouse is 48 inches, and 2 inches are allowed for clearance on each wheel, it equals 52 inches. The tire dimension being 4 inches, this will equal 2 inches on each side, making 2+2+52=56. Now, if the wheelhouse is 48 inches,

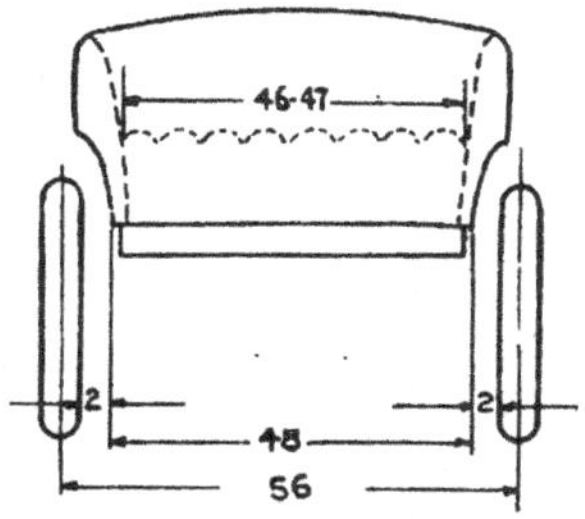

FIG. 41
Rear end of body showing its relation to rear wheels

the best width possible is between 46 and 47 inches. This only allows from ½-inch to 1 inch on each side for trimming. If the cushion is raised above the wheelhouse, it makes a very ungraceful looking seat, which is not satisfactory with the present style of bodies. Then, again, the cushions will have to have the extensions run-

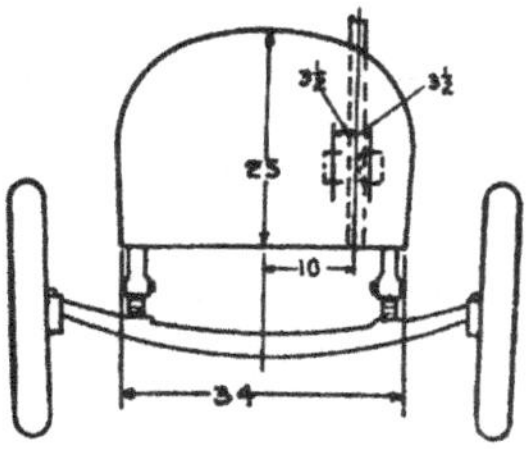

FIG. 42
Average condition obtained at the dash

ning out, and these would not be deep enough to give riding comfort. A depth of 8 to 10 inches is given on the cushion springs in order to have a comfortable seat. From the foregoing the reasons for the 46-7 dimensions are explained.

The diagram shown in Fig. 42 illustrates the average condition obtained at the dash. The width of the dash must be sufficient to allow ample clearance for the feet when operating the pedals. The modern tendency is to have a wide dash, and the figures given are average ones, such as provide clearance.

CHAPTER IV

CHASSIS PARTS TO BE CONSIDERED IN BODY DESIGNING

The first consideration for the body engineer when making a design for a body is to find out the details of the chassis upon which the body is to be assembled. Most car builders furnish chassis blue prints which have practically all the information required by the body builder.

If the request calls for a body design alone the car manufacturer will be expected to provide body hood line, front and rear fender, running board location, wheelhouse clearance and also frame clearance. If a new car is to be designed, in all probability the radiator, hood, dash shape, front and rear fender, sides, aprons and running board location have to be designed and located by the body engineer. In the latter case the body engineer will require the wheelbase, tire size, height of frame from the ground and the shape of frame, and the dash position, which, of course, will be located so as to give enough room for the engine.

Fig. 43 shows a chart that has the information necessary for a "style drawing," A represents the wheelbase or distance between the center of the front wheel and the center of the rear wheel. B is the diameter of the front tire. B 2 is the diameter of the rear tire. C is the height from the ground to the top of the frame when the car is carrying the complete load that it is designed to carry. D is the height from the ground to the top of the frame at the rear of the seat with the car completely loaded. E is the distance from the front of the frame to the center of the wheel. F is the distance from the center of the wheel to the front of the dash. G is the distance from the dash to the back of the wheel. H is the height from the top of the frame to the under side of the steering wheel. I is the diameter of the steering wheel. The front pedals should be located if the seats' backs are to be correctly placed, and dimension J will give this.

With the information given in this chart, accurate style drawings can be made. It very often happens that the information suggested above is not available and the body engineer has to use his own judgment in making the preliminary sketches. When the style of the body has been determined, further information will be required, starting in with the frame. This should be laid in on the body drawing, and all cross members, rivets and all other projections liable to cause body interference carefully inserted on the drawing.

Automobile chassis frames vary considerably, as shown in Fig. 43-A. There

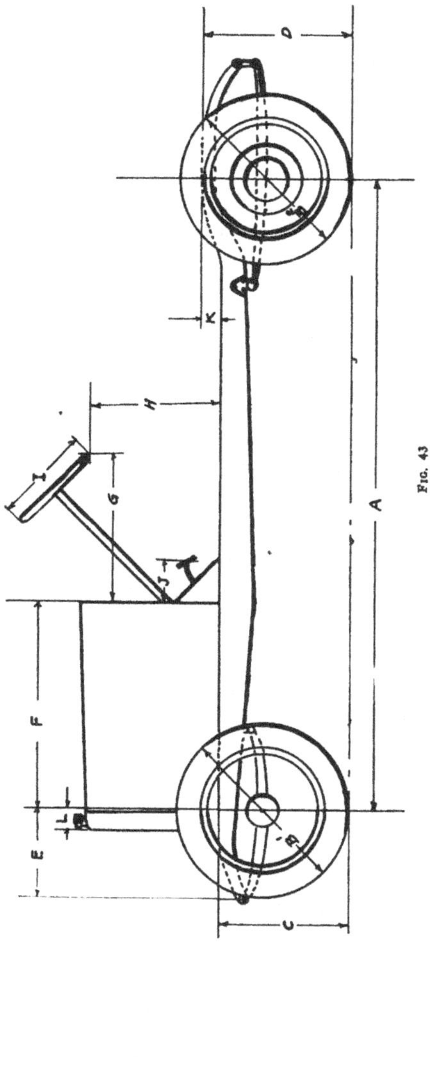

Fig. 43
Chassis chart giving the information necessary for a body builder's "style drawing."

are three general types of frames, 1st: Flat frame with parallel side rails, 2nd: Flat frames with the sides rails tapering from the back to the front, 3rd: Frames with the "kick-up" in the rear, either taper of parallel side rails. The first class

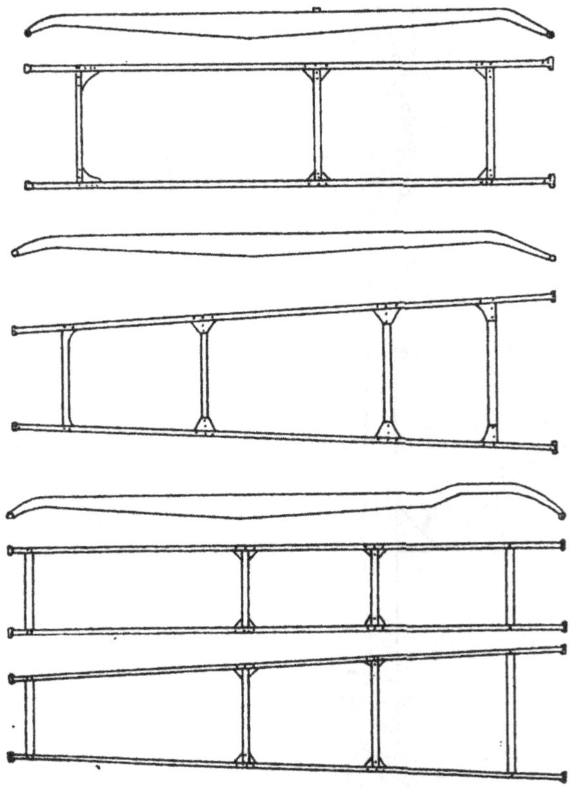

FIG. 43 A
Automobile chassis frames vary considerably as shown in these top and side views

of frames permit of making flat sills and simplify the body construction. The second class permits the use of flat sills, and also makes it possible to have a narrower sill than the first type of frame. The frames that have "kick-up" require a built-up body sill. The kick-up is provided to give axle clearance.

Starting from the front of the body the details which have to be considered in a body layout will be taken up. The dash must be located to give ample clearance from the engine and any of its component parts, allowing for any possible

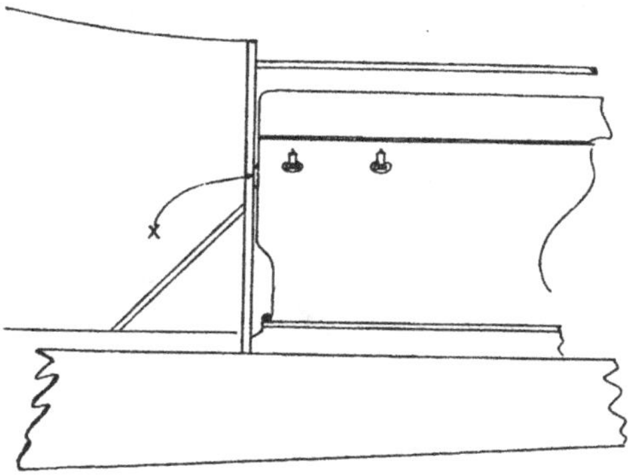

FIG. 44
An example of interference between dash and engine

manufacturing variations. It is a common fault to try to get all the body room possible and then have trouble with engine interference. See Fig. 44.

The toe boards are laid out to allow the proper clutch and brake pedal travel, and the sides of the shrouds must give clearance for the pedal pad. (See Figs.

FIG. 45
The toe boards are laid out to allow proper clutch and brake pedal movement

45 and 46). The toe and floor boards must also clear the engine housing and transmission as indicated at X in Fig. 45. In making up the toe and floor boards suitable cut outs have to be made for the foot pedals, starter pedals, accelerator

pedal, speedometer cable, steering column, gear shift lever and the emergency brake levers.

Sometimes the battery is located in the body, but generally it is supported in the frame. There are several locations for this unit. It can be set in the right or left side of the frame near the front seat heel board or in the rear under the

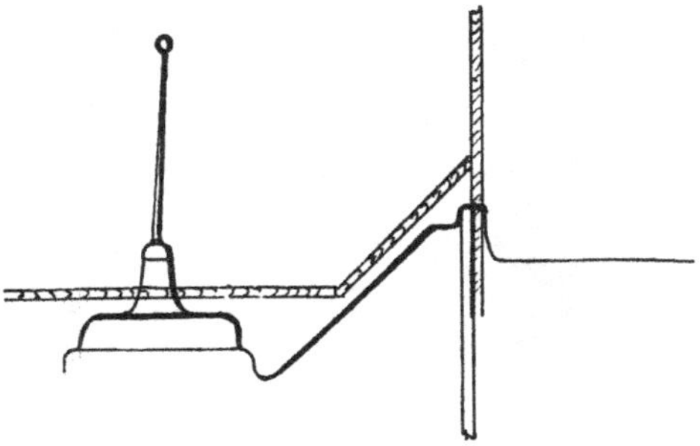

FIG. 46
Suitable provision must be made for all controling levers

tonneau floor boards; in the left or right side of the frame. A good location is on the outside of the frame, between the running board and the frame, under the side apron. As a last resort it can be carried on the top of the running board. The point to be made is that the battery location must be determined so that suitable provision is made for it and the body construction carried out accordingly. When the general outline of the body has been laid out, the position of the body

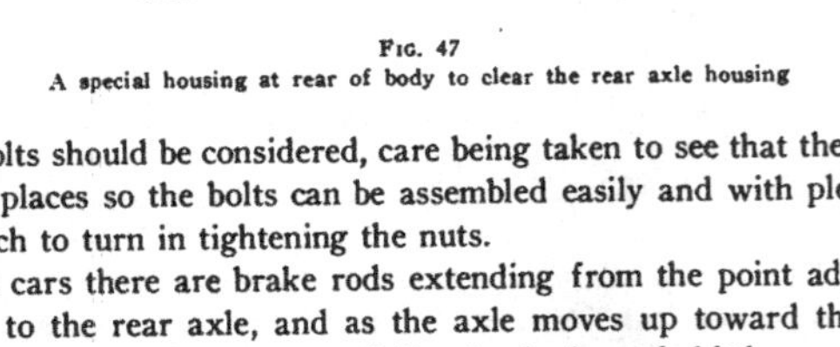

FIG. 47
A special housing at rear of body to clear the rear axle housing

hold-down bolts should be considered, care being taken to see that they are located in accessible places so the bolts can be assembled easily and with plenty of room for the wrench to turn in tightening the nuts.

In most cars there are brake rods extending from the point adjacent to the transmission to the rear axle, and as the axle moves up toward the body, care must be taken to see there is no possibility for brake rods hitting any part of the body when the springs deflect. It is very often necessary to make a special housing at the rear of the body to clear the rear axle housing, as shown in Fig. 47. The size of this housing varies, so it must be considered on all bodies, provision being made for ample clearance for the part when the axle is up as high as it can possibly go.

In some cars the gasoline tank is in the frame or under the front seat. This must be located. However, in most of the modern cars the gas tank is in the rear and hung between the side rails on the frame. The gasoline tank has gauges and filler cap. These must be accessible so the body must not extend back so far that it will interfere with them. There are always some lighting wires which have to be attached or fastened to the under side of the body, and the tail lamps wire at the rear generally require a cut-out in the rear sill.

CHAPTER V

RADIATOR DESIGN

The radiator, as well as being one of the important factors that determine the lines of the car, is one of the main mechanical parts, too. Its function on the chassis should be clearly understood, so that in contemplating a design the factors that govern the size and shape of the radiator will be properly considered. On the earlier design of car the radiator was arranged to comply with the mechanical requirements, and a shape was laid out that was considered suitable. Then the chassis was turned over to the body designer to put the body on.

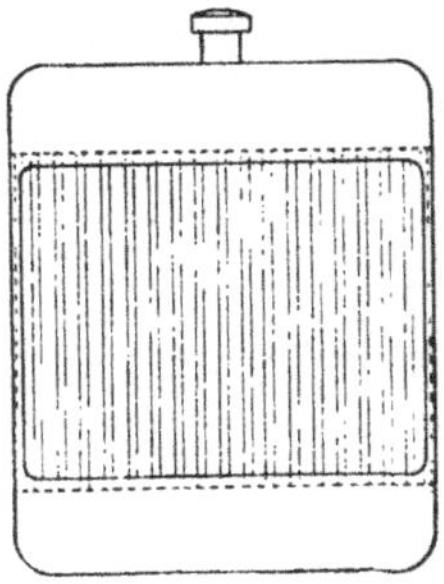

FIG. 48
The mechanically ideal shape for a radiator would be rectangular or square

With the demand for improvements this condition has been changed, and in most cases the radiator shape is considered in its relation to the general design required. As the present-day automobile is the sum total of all its parts, to get the final result it is readily seen how essential it is to have a thorough understanding as to what can be done with the radiator before the new ideas are applied. There are a few cars yet that adhere to their original design of radiator. Some are retained because their original patterns happen to work out very well in the latest design, but the majority are retained because they are known trade marks of certain cars and therefore possess considerable advertising value. The object of this chapter is to explain the basic principle upon which the radiator design depends, and then to review the shapes and designs for the different types of radiators, past and present. If the existing shapes are known, the new designs can be worked out, basing their shapes on the designer's ideas.

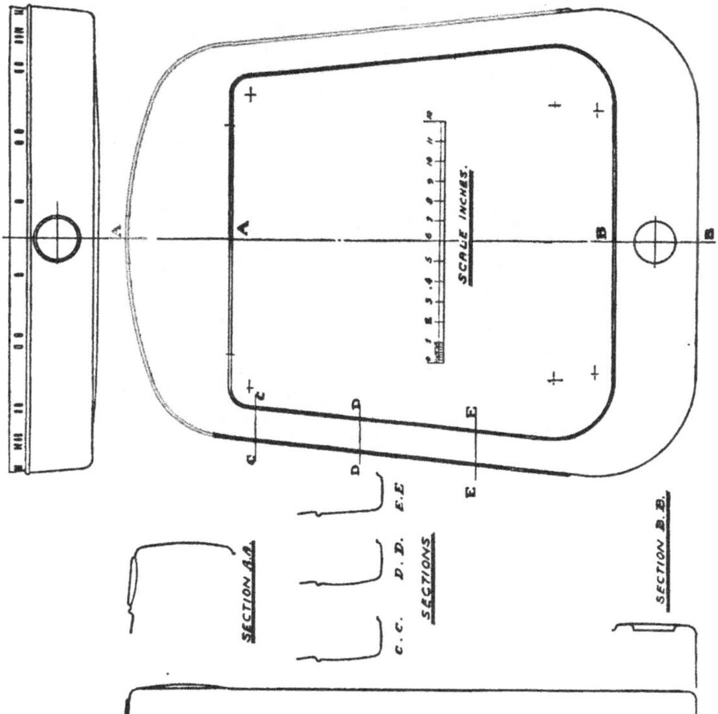

FIG. 51

Drawing illustrating a good design for a radiator shell. The top and bottom have easy curves and the radii at the corners are large, the sides alone being straight.

When the various shapes are grouped together as shown in this chapter, it will be seen that almost every possible variation of shape has already been tried, and it will at once be appreciated how necessary it is to have the different shapes for reference and for study, as it will require a very thorough familiarity with existing shapes before a new and original shape can be created.

Before starting a design for a radiator the mechanical and manufacturing limitations should be considered, because it would be of no use to create a beautiful design for a radiator if it could not be made, or if it would not function properly. It is not within the scope of this book to go into all the mechanical and scientific values of a radiator. Only the fundamental parts of the radiator need be considered in order to analyze the possibilities of designing or developing its general external appearance.

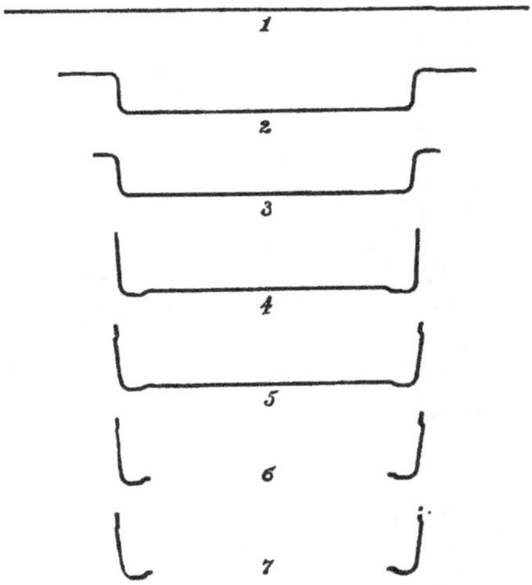

FIG. 53
Difference in the shape of a shell after each of seven operations

In order to lay out a radiator, the heat-dissipating requirements of the engine it is to be used with must be known. Then the type of core must be considered, as there are a number of different styles of cores and they have different relative efficiencies. Having decided upon the type of core, and its efficiency being known, the frontal area can be determined. If the frontal area required by the core is too large, it may be possible to reduce the area by increasing the depth, or, *vice versa*, the depth may be decreased to obtain a larger front. The larger front is the most desirable, as the efficiency per weight of material is higher than in the small front

Fig 49

Fig 50

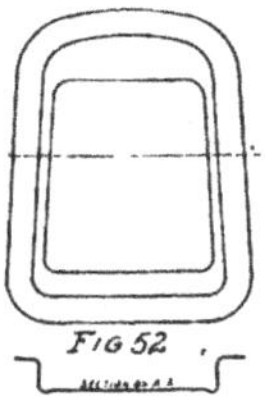

Fig 52

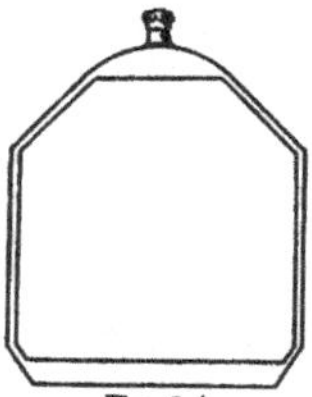

Fig 54

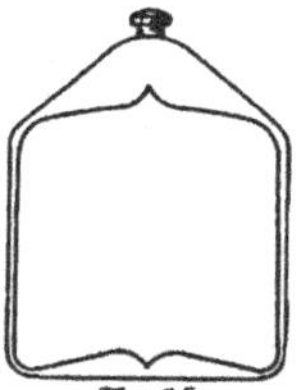

Fig 55.

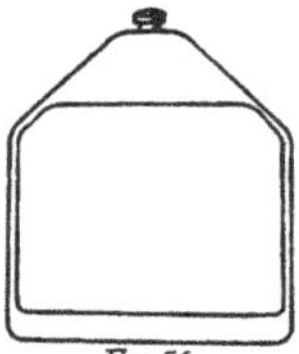

Fig 56

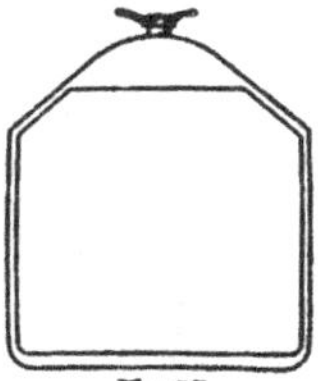

Fig 57

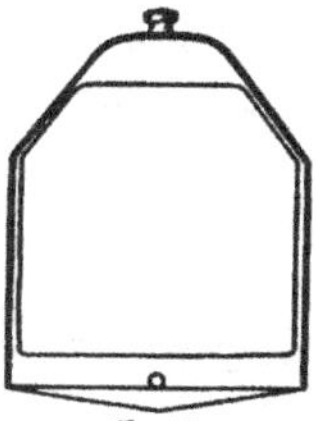

Fig 58

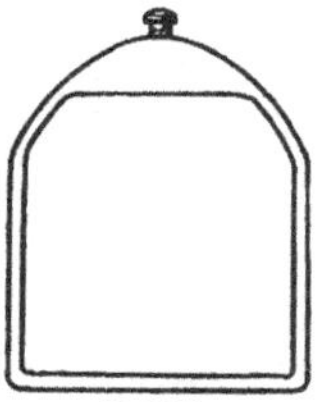

Fig 59

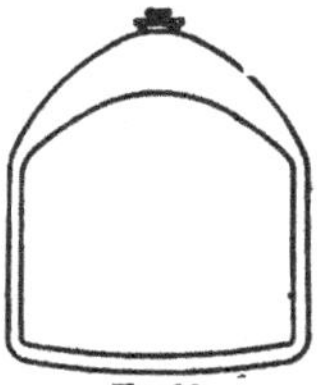

Fig 60

Fig 61

Fig 62

Fig 63.

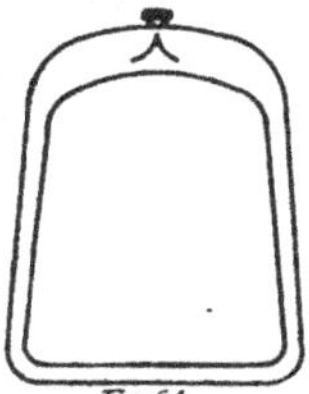

Fig 64

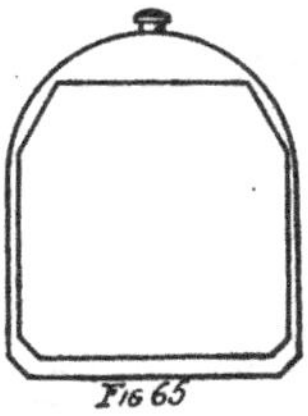

Fig 65

SHAPES AND DESIGNS OF AUTOMOBILE RADIATOR FRONTS—PAST AND PRESENT

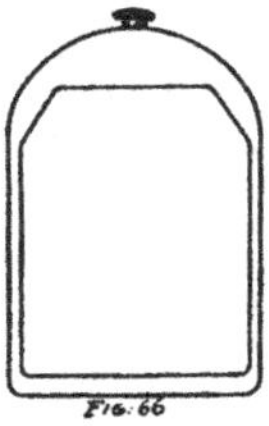
Fig. 66

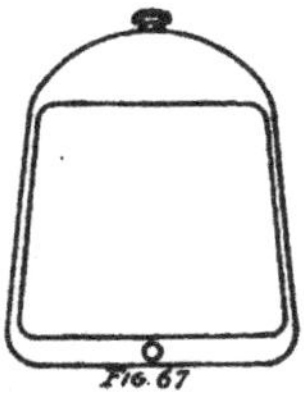
Fig. 67

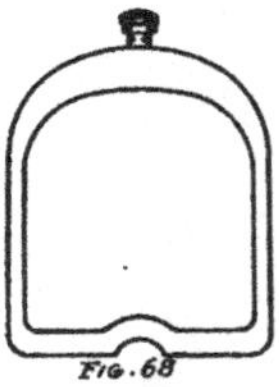
Fig. 68

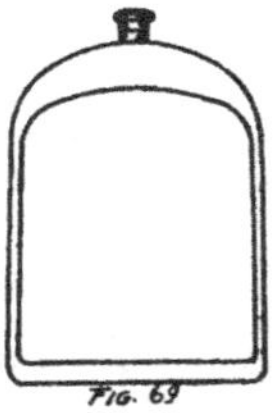
Fig. 69

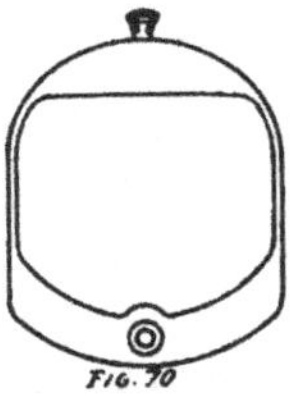
Fig. 70

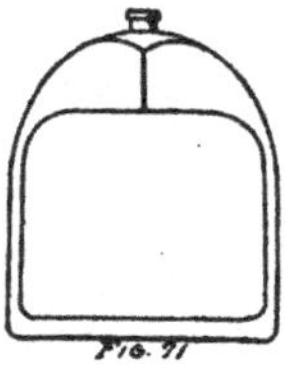
Fig. 71

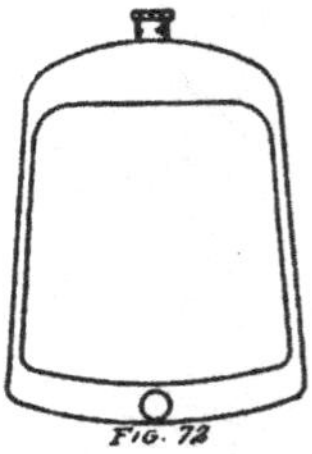
Fig. 72

Fig. 73

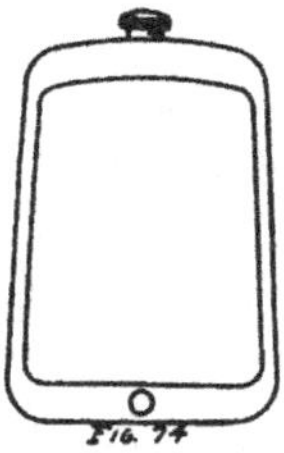
Fig. 74

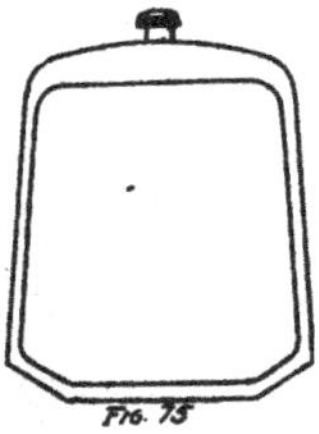
Fig. 75

Fig. 76

Fig. 77

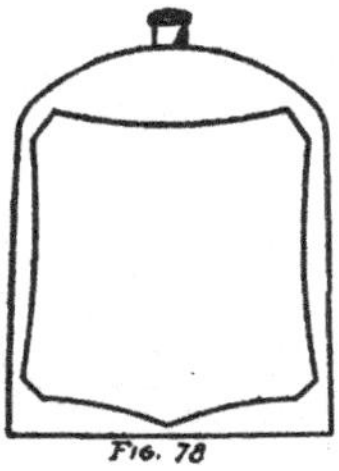
Fig. 78

Fig. 79

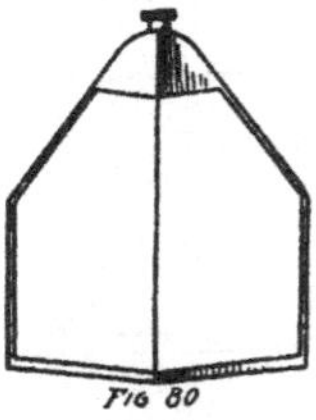
Fig. 80

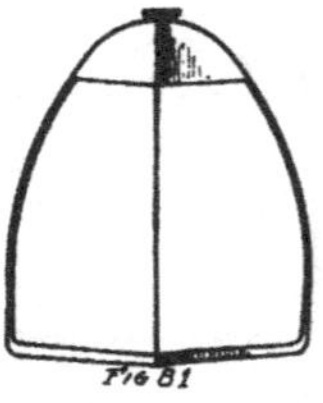
Fig. 81

SHAPES AND DESIGNS OF AUTOMOBILE RADIATOR FRONTS—PAST AND PRESENT

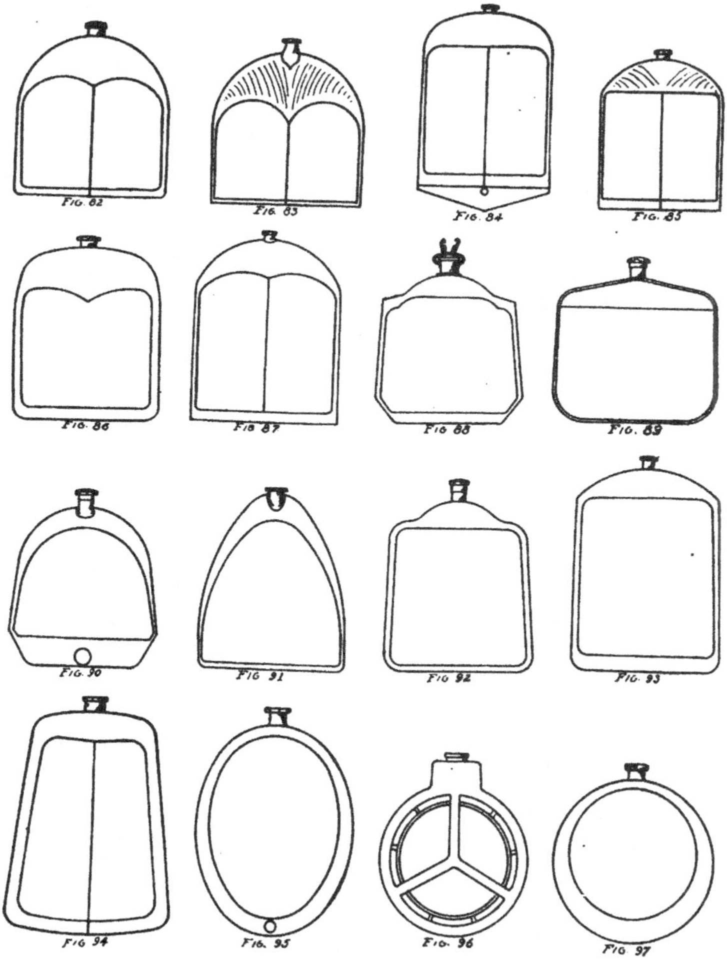

SHAPES AND DESIGNS OF AUTOMOBILE RADIATOR FRONTS—PAST AND PRESENT

type with a greater depth. The size of the core being decided upon, the top and bottom tanks must next receive attention.

The capacities of the tanks must be within certain limits. An engine with a circulating pump can do with smaller tanks in the radiator than an engine that depends on natural circulation, or what is called the thermo-syphon system.

The mechanical construction of a radiator can be divided into five main parts, viz., the core, the upper tank, the lower tank, the shell, and the brackets that hold the shell on to the other portions. The two tanks are connected by the core, which is composed of a number of tubes with radiating fins of a tubular construction that will present the greatest possible surface to the air, so that the heat of the water flowing through it will be rapidly dissipated. The tanks and core are contained in the shell, and the unit as a whole is suitably attached to the car frame.

The mechanically ideal shape for a radiator would be rectangular or square. This would give the maximum capacity per weight and be the most simple to construct. In Fig. 48 (page 38) is shown a simple rectangular-shaped radiator.

It is important to see that the radiator is designed so that it will not unduly constrict the effective size of the core, and also to see that it can be made as simple as possible. To illustrate how the core's efficiency is interfered with, two radiators are shown in Figs. 49 and 50 that constrict the upper tanks so that extensions are necessary in order to get an effective number of tubes in it. It is readily seen that the cost of manufacture is greatly increased by these designs, and the number of tubes is reduced and the efficiency of the radiator correspondingly impaired.

Having studied the inner workings of the radiator, the outside or the shell can now be taken up. When the factors that govern the inside of the radiator are known, it is possible to make the necessary compromise to obtain the desired external design in conjunction with the required efficiency. The outer shell of the modern radiator is made of steel, and is formed to shape in dies. In the case of small quantities it is made of sections joined together, but as the production is so large even on the average cars, dies are generally made to produce the entire shell from one piece of steel. This method is by far the most satisfactory and the most economical one from the maker's point of view. But, to make a shell from one piece requires an immense press and large dies, and then in order to make the shells successfully the design must be governed by certain conditions that are required to make the dies work properly.

To obtain the best results in drawing a radiator shell the design should be one that will give a uniform draw all around. The best shape for an ideal draw in the dies would, of course, be circular, but a circular shell would not give an economical condition for the tanks or the best possible efficiency of the radiator; so, speaking for the average shell, the design will be rectangular with an easy running top, and a full radius will be allowed at each corner. The front edge of the shell will have a radius; the sides will be two or three degrees tapering. This will allow the shell to come off the punch easily, and also come out of the die readily.

In Fig. 51 is shown a drawing of a good design of radiator shell. Upon looking at the front view it will be seen that the top and bottom have good, easy curves, and the radii at the corners are large, with the sides alone being flat. In the top view is shown how the sides bevel off to the front. The section shown at A A, Fig. 52, shows raised portion on the top of the shell for the radiator filler base to go through. This raised portion has a uniform height from the plane of the shell's taper on top.

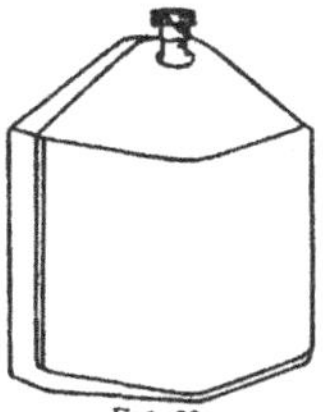

FIG 98

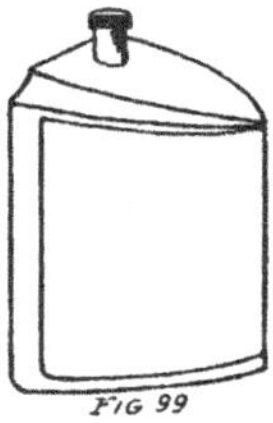

FIG 99

FIG 100

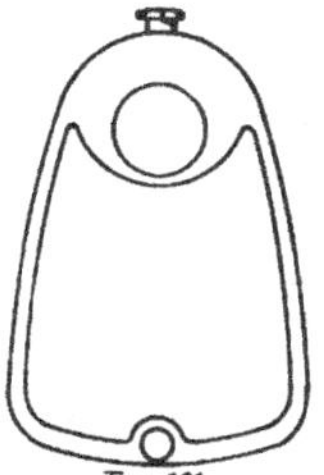

FIG. 101

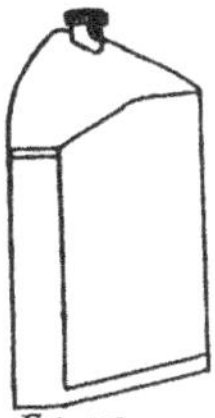

FIG 102

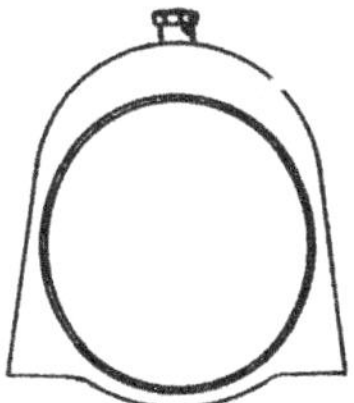

FIG. 103

Shapes and designs of automobile radiator fronts—past and present

The crowned front has an easy curve, the object of this being to stretch the metal so that there will not be any wrinkles in it, and also to set the metal so that it will retain the desired shape. For the same reason all surfaces of the shell have slight crowns in them. For instance, the section B B shows the crown at the bottom of the shell. To make a shell similar to the design shown in Fig. 51 it takes seven operations; these are as follows: No. 1 blank, No. 2 draw, No. 3 trim, No. 4 second draw, No. 5 expand, No. 6 take out front, No. 7 pierce the holes in sides and top.

In sketch No. 52 is shown what the shell looks like on the second operation, and in the sections, Fig. 53 (numerals from 1 to 7) are shown the difference in the shape of the shell after each stage of its seven operations. When the fundamental principles of the radiator's construction are understood a design can be worked out that they will prove satisfactory, providing the different points are kept in mind and the design made accordingly. The primary condition having been studied, the next problem is what shall the shape be, and in order to assist the choice of the shape the various styles of radiators have been grouped together and are shown in sketches from 54 to 103, beginning on page 41.

Eight groups have been made. The first group has been called the angular top; the second, the flat top; third, round top; fourth, flatted round top; fifth, V-fronts; sixth, miscellaneous; seventh, round and oval shaped fronts; eighth, miscellaneous foreign types. Group 5, V-fronts, could be subdivided again into two groups, the first to have sharp V-fronts and flat shaped tops, and the second to have round top with V-fronts. As far as the front view is concerned, the shapes are of all the types grouped from 1 to 8. The sketches show fifty different types, and each one has some characteristic that the others have not, and, while these do not show every type of radiator made, they show the great range through which the designers have worked to get new and desirable designs.

In group No. 1 are shown sketches of the Pierce-Arrow; Fig. 54, Fig. 55, National; Fig. 56, Studebaker; Fig. 57, Stutz; Fig. 58, Chevrolet; Fig. 59, Cadillac; Fig. 60, Overland. While they are all of the same general outline, not one of them could be mistaken for the other.

In group No. 2, the Peerless, Fig. 61, has an angular side, but also has a flat top, while the Ford, in Fig. 62, has a flat, shooting top, and the Murray, in Fig. 63, has an even more decided flat, sharp-cornered top. This shape is similar to a number of English cars, viz., Rolls-Royce, Lanchester, Edge, Armstrong.

Group No. 3 is a rounding top class, and, starting with the Marmon, Fig. 69, with not quite a complete radius at the top, and then the Kissell, Fig. 65; Mitchell. Fig. 66; Buick, Fig. 67; Wescott, Fig. 68; Empire, Fig. 69; Haynes, Fig. 70; and Reo, Fig. 71; these all have decidely rounded tops. The Reo, in Fig. 71, has a small, sharp point at the top only, and could not be termed a V-front.

Group No. 4 comprises the Hudson, Fig. 72; Mercer, Fig. 73; Chalmers, Fig. 74; Chandler, Fig. 75; Dodge, Fig. 76; Hupmobile, Fig. 77; King, Fig. 78; and the Saxon, Fig. 79. Each one in this group can be easily recognized by its characteristic.

In group No. 5 are the V-fronts. The first one is the Owen Magnetic, Fig. 80; next is the Singer, Fig. 81. These are the only ones with the "V" portion starting at the extreme top. In Fig. 82 the Apperson has a top rounding over to the V. The Moline-Knight, Fig. 83, is similar; so also is the Paige, Fig. 84. The Scripps-Booth, Fig. 85, has a flat portion going to the V. The Maxwell, Fig. 86; is not strictly a V, as the top portion alone is crowned V-like. The Pathfinder, Fig. 87; has a decided V-front.

In group No. 6, miscellaneous, comes several different characteristic designs.

Fig. 88 is the Packard; Fig. 89, the Winton; Fig. 90, Olds; Fig. 91, Novara; Fig. 92, White; Fig. 93, the Liberty; Fig. 94, Bour-Davis.

Group No. 7 has not many adherents. The Phianna, Fig. 95; the Buss type, Fig. 96; and the Delauney-Belleville, Fig. 97; are the only exponents in this country.

In the foreign groups are shown some radical designs to illustrate some more phases in the trend of radical design. Fig. 98 is the Belgian Metallurgique. Figs. 99, 100, 101, 102 and 103 are various types used on European cars, and show a wide range in ideas. No attempt will be made to create a new design; this will be left to the designer to interest himself in, as the collection of shapes in this book will illustrate very aptly what there is before the man who desires something new and original in radiator shapes upon which a new body line can be developed.

HOODS, HOOD SILLS AND HOOD FASTENERS

The automobile hood primarily forms a cover for the engine and its parts, and in order to do this it has to extend from the radiator to the front end of the body. Suitable ledges are provided on the radiator and the body for the hood to rest upon. The hood in its simplest practical form consists of four main parts, two upper sections right and left, and two lower sections right and left. The four pieces are hinged where they attach to one another so that the hood may be easily

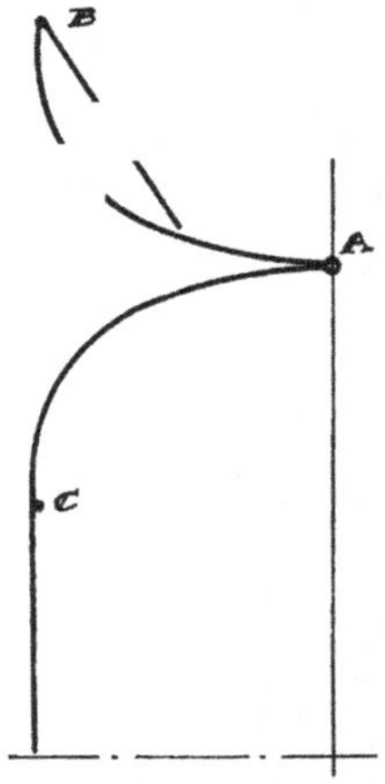

FIG. 104
The four sections indicated, with the right side lifted up. Note hinges at A, B and C

lifted up from either side of the car and folded up so that there is ready access to the engine. See Fig. 104, which shows the four sections with the right side lifted up and note that it is hinged at A, B and C.

It is generally necessary to provide ventilators to allow the heated air to get out readily all along the sides or at the ends or on top as the case may be. The ventilators are generally located in the lower side sections, but sometimes they are put in the upper section or a plain centerpiece is provided at the top into which ventilators are set. This makes it necessary to add a fifth section to the hood and an extra hinge. In Figs. 105 to 108 are shown four different styles of "louvres," as these ventilators are called. Fig. 105 is the prevailing type. Fig. 106 is located at the top and, while efficient, is expensive to make. Fig. 107 is located in the top section and Fig. 108 is a circular-shaped louvre on the sides.

The hood is generally made of steel, 22 or 20 USS gauge thick, but aluminum is used on some high-priced cars, the gain being a saving of a little weight. When aluminum is used, steel reinforcements have to be used while a sheet steel hood is satisfactory without reinforcements. In Fig. 109 is shown a scale drawing of a

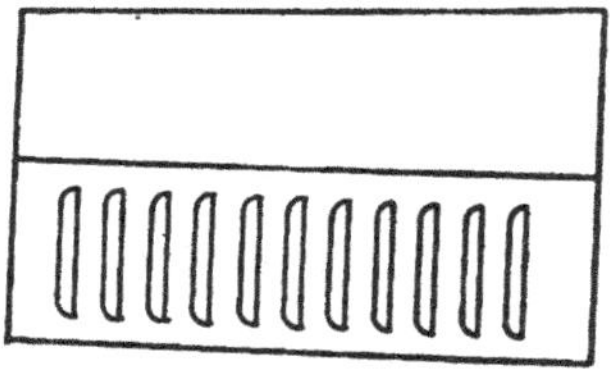

Fig. 105

Fig. 106

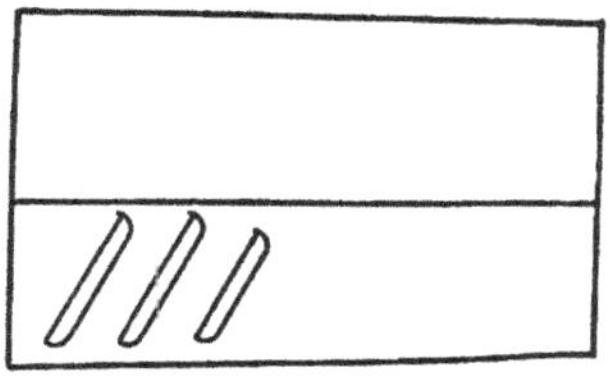

Fig. 107

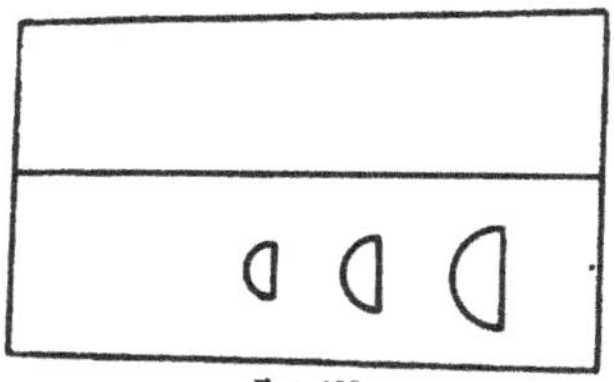

Fig. 108

In Figs. 105 to 108 are shown four different styles of ventilating slits or "louvres," as they are properly termed

hood that is patterned after a very successful design. Upon examination it will be seen that the front end is lower than the back and that the front is also narrower. This is a prevailing condition, as the body designs demand this tapering effect. The top section is made in a die which forms the cone-like taper in it, and at the same time turns up the hinge section at the top and bottom. The sides are practically straight and flat; the back end of the side hinge is located ⅜-inch higher than the front, so that it will appear straight to the eye. If it were actually level it would appear that the back were lower than the front, owing to the hood being wider at the back than at the front. At the rear end of the hood a bead is formed to stiffen it and make a finish on the end (see section CC). The lower sections have the ends turned over and folded down flat to stiffen them (see section AA). The top hinge sections as shown in the drawing (section BB) is inter-

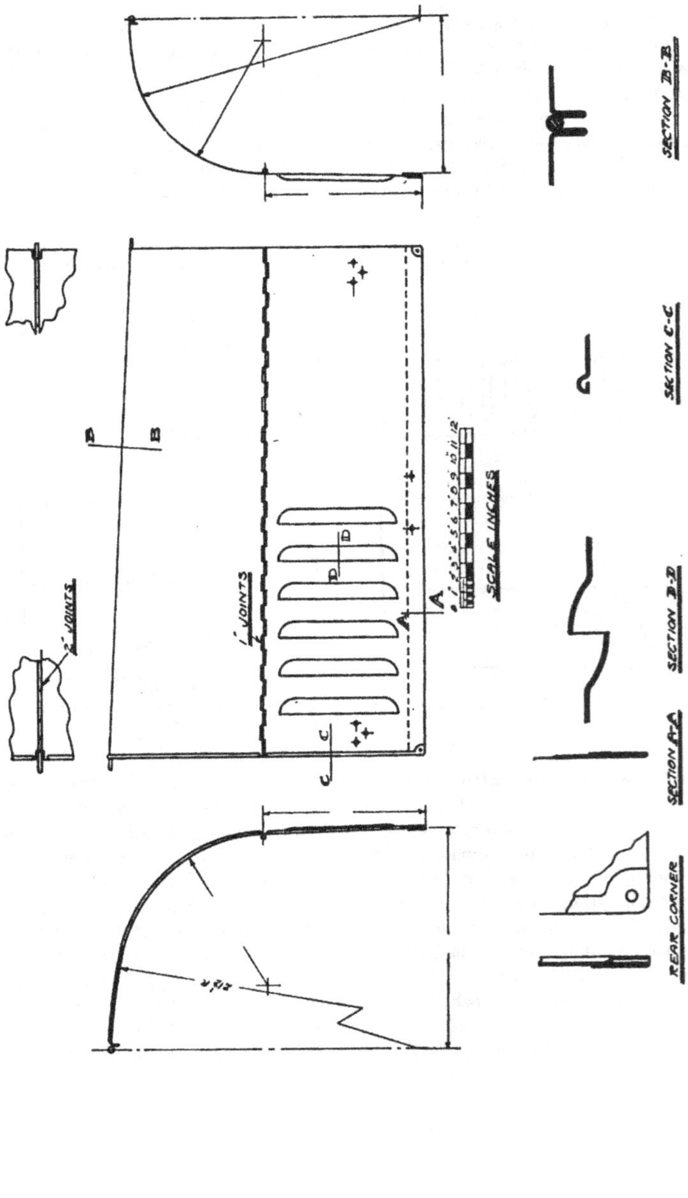

FIG. 109

Scale drawing of a hood that is patterned after a successfully designed car esting because it is formed up from the hood and provides a stiffening rib as well as a hinge. The lower hinge is formed right out of the metal too, and is a very simple construction. The top hinge uses a 2-inch joint while the lower requires a 1-inch. The small top view shows the cutouts that are necessary for the clip that secures the hood to the body and radiator.

In section DD is shown how the louvres are depressed in the hood. One-half the depression is on the outside of the hood and the other half in the inside, the object of this is to get a maximum opening without having an unsightly projection on the outside. In the sectional view shown at the rear corner a depression is illustrated; this is for a leather or rubber washer which is attached to the corner of the hood to prevent the hood scratching the paint off when it is let down hurriedly and might strike on the shroud. The holes shown in the side of the hood are for the handle and hood fastener brackets.

Fig. 110

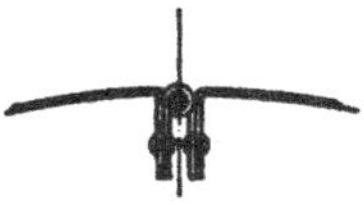

Fig. 111

Fig. 112

Fig. 113

Fig. 114

Fig. 115

Fig. 116

Fig. 117

Fig. 118

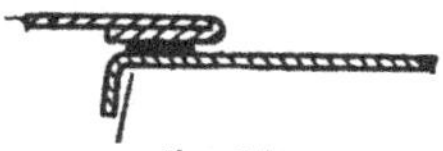

Fig. 119

Various types of hood hinges, beads and methods of fitting the hood to the body

In designing a hood the hinges and the cutouts must all allow the paint or enamel to be readily applied. The majority of the medium-priced car makers enamel the hoods and it is necessary to have as few projections and holes as possible so that the enamel will not collect and also make dirt runs. When the hoods are enameled all attachments, such as hood fasteners, brackets and handles, should be applied after the hood is finished, as the workmen could not enamel the hood with these parts in place. The rods used in the hinges are rust-proofed by Sheradizing or Parkerizing so that they will always work freely.

While the foregoing outlines a good standard design of detail construction the various other designs will be interesting to review. Starting with the top hinge the earliest type of hood hinge was a piano type, which was riveted to the two sections of the hood as shown in Fig. 110. This, it will be seen, projects above the hood line and makes an unsightly appearance, and as the car lines are desired smooth as possible this is not satisfactory.

In order to get the hinge flush a method was worked out as shown in Fig. 111. The hood section is formed down and the hinge is riveted to it; as this makes the hood stiffer a smaller rod can be used in the hinge. Very similar results are accomplished by making the hinge like that shown in the hood drawing, Fig. 109,

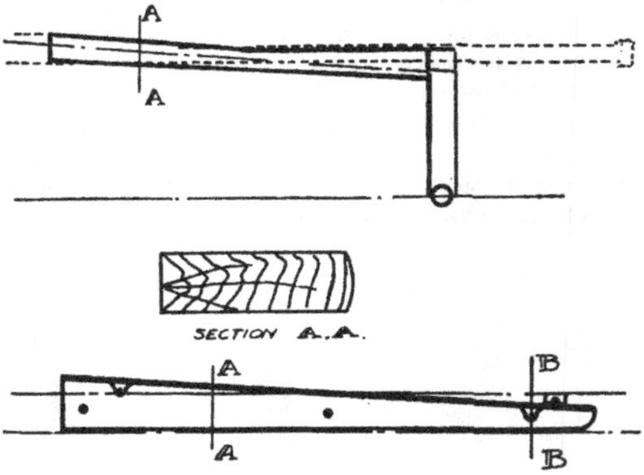

Fig. 120
General shape of wood hood sill

and at the same time at a greatly reduced cost. The side hinges, constructed as shown in the drawing, Fig. 109, are of the simplest design possible, but they have two criticisms. One is that they are not water tight, and the other is that they show the notches. In Fig. 112 is shown a section of a hinge that is used a great

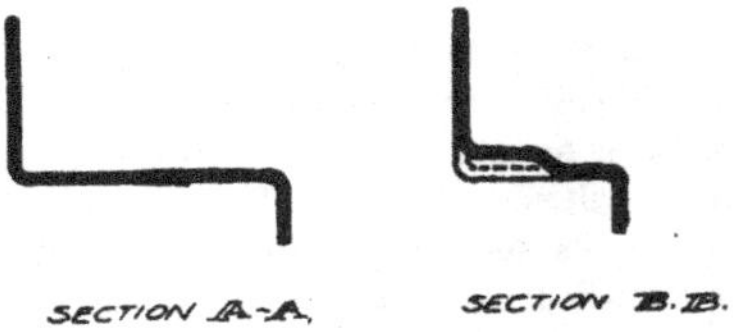

Fig. 121
Sectional views on A-A and B-B of above diagram

deal, and this makes a neat appearance and is water tight. Another way of getting this result is shown in Fig. 113. A piano type hinge is used and securely locked in each section of the hood as shown.

There are several different methods of putting a bead or finish on the front or rear end of the hood. In Fig. 114 the end of this metal is turned over on to itself; in Fig. 115 a molding is pressed in. With this style the edge is liable to catch on the body, so the section CC shown in the hood drawing, Fig. 109 is better. In Fig. 116 the bead is set up so that it will not catch on the body and also to provide more clearance at the body. In connection with these moldings the method of fitting the hood has a great deal to do with the results they give, for instance there are three principal ways of fitting the hood on the body. The first, shown in Fig. 117, provides a depression in the cowl for the hood to rest on. With this style the hood finish shown in Fig. 109 is adapted. The second method, shown in Fig. 118, provides an aluminum strip or molding, against which the hood fits. With this style the hood finish shown in Fig. 114 or Fig. 115 is adaptable. The third method is to let the hood rest on top of the cowl, which does not have a depression, as shown in Fig. 119. With this method the hood finish shown in Figs. 114, 115 or Fig. 116 can be used. It will be noted that a belt or lace is set

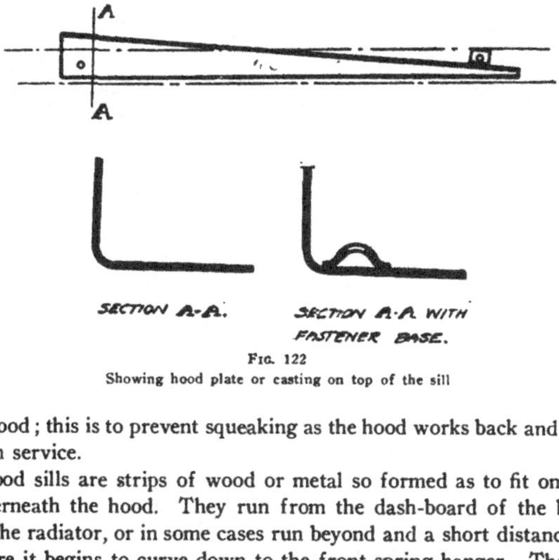

Fig. 122
Showing hood plate or casting on top of the sill

under the hood; this is to prevent squeaking as the hood works back and forth when the car is in service.

The hood sills are strips of wood or metal so formed as to fit on top of the frame underneath the hood. They run from the dash-board of the body to the bracket of the radiator, or in some cases run beyond and a short distance down the frame, where it begins to curve down to the front spring hanger. These sills are attached to the frame and make a good finish and also a convenient place to attach the hood fasteners.

There are three general designs worth considering, the first being the one made of wood. The general shape of a hood sill of this type is shown in Fig. 120. It will be seen that at the body end the sill projects to the outside of the frame while on the radiator end the sill projects to the inside of the frame. The hood line is shown dotted in and this is what determines the shape of the hood sill to the greatest extent.

The hood sill shown in Fig. 121 is the second type and is made of steel formed up as shown in the section AA. In section BB is shown the depression made for the hood fastener. This saves the addition of a hood plate or casting on top of the sill as is necessary in the third style shown in Fig. 122. The general shapes are identical, in each case the sections alone being different. In the last case the

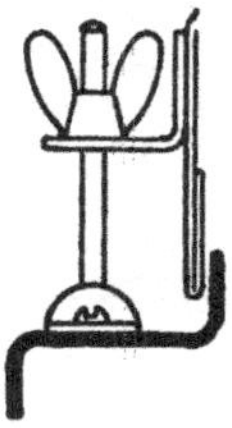

Fig. 123
Bolt and wing nut fastener

Fig. 124
Fastener hooking on to a bracket

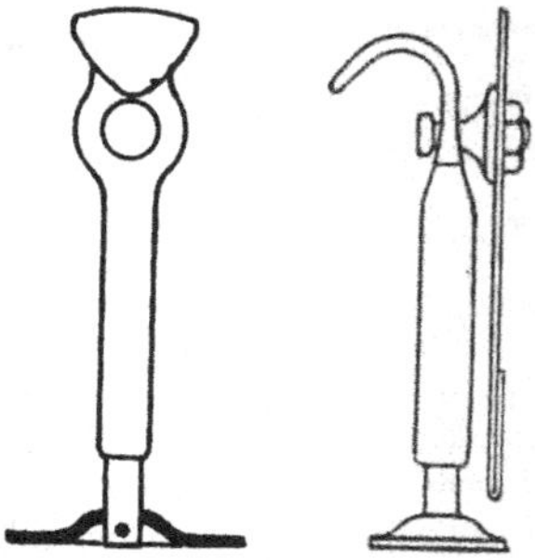

Fig. 125
Fastener hooking on to a stud

section is just a plain angle, and obviously this is a cheap construction but does not give the finished appearance of the second type, although it is very extensively used.

The hood fastener is the device used to secure the hood to the car. The hood must be held down well and so clamped that it does not rattle. There are several different ways of doing this, and a few of the fasteners that are in most general use will be described.

In Fig. 123 is shown a bolt and wing nut fastener. The sketch is self-explanatory—to relieve the hood the wing nut is turned up. There is a serious ob-

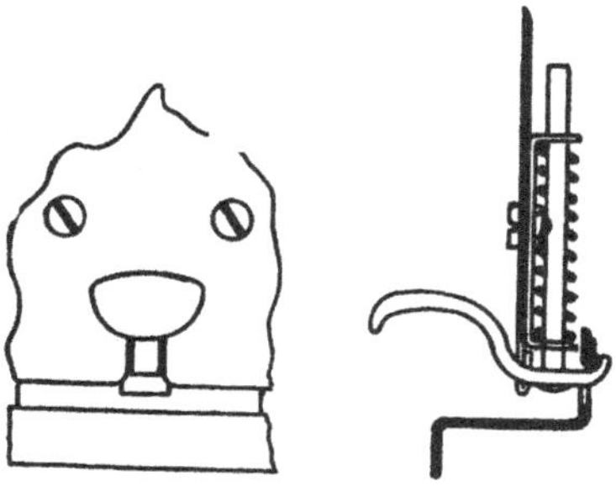

FIG. 126
Concealed type of fastener in which only the handle is visible from the outside

FIG. 127
Another type of concealed hood fastener. Ferro Stamping and Mfg. Co.

jection to clamping the hood down rigidly because when the car is in service the frame distorts, which tends to throw a great strain on some parts of the hood. The "hook-on" fastener is perhaps the most used type of any at the present time and it is made in several different detail variations.

In Fig. 124 and Fig. 125 are shown two of these types of fasteners. One hooks on to a bracket and the other on to a stud. The principle upon which these work is as follows: When the handle is raised it compresses the spring and the barrel, and the bracket is so set on the hood that the fastener has to be raised to hook it on, and when this is hooked on the tension of the spring tends to pull the hood down and keep it in position.

In Fig. 126 is shown a concealed type of fastener which is a very neat design, as only the handle is visible from the outside. This same handle obviates the necessity of a hood handle for raising the hood. When the fastener is in position it is hooked on to the sill and the spring is composed so that it tends to pull the hood down and hold it securely.

In Fig. 127 is shown another concealed type of hood fastener. This consists of brackets attached to the cowl and radiator. These brackets have flat springs in them so they will engage with the rod in handle section, which is attached to the hood.

CHAPTER VII

TOURING CAR BODY DESIGN

The general types of bodies and the principles upon which the main dimensions are developed have been shown in the previous chapters. The various details that have to do mostly with the general appearance, capacity and convenience of a body will now be gone into.

A touring car may be any sized car and have any capacity from two upwards, but the word touring car is accepted usually as meaning either a five or seven-passenger car. There are cars in the market which provide room for from four to six or seven passengers. There is very little difference in the general design between a 4, 5, 6 and 7-passenger car. Some manufacturers give a seating arrangement at the rear for two passengers instead of three. This permits them to make the body a little narrower at the rear than it would be with 5 or 7 passengers.

It was shown in the preceding chapter that the wheels require a definite clearance, consequently the best possible arrangement of the rear seat cannot make it really comfortable for three extra heavy people. In making up the seat for two only, ample room can be allowed for this seat. In either the 6 or 7-passenger car the two extra seats are located between the back of the front seat and the rear seat. These extra seats are generally folded up when not in service. The seating capacity being determined, the number of doors required and their location should be considered.

There are eleven different ways to provide means for getting in and out of a touring car, and these are as follows:

(1) Two doors—one front right hand; one rear right hand.
(2) Two doors—one front left hand; one rear left hand.
(3) Two doors—one front left hand; one rear right hand.
(4) Two doors—one front right hand; one rear left hand.
(5) Two doors—two rear doors with a front divided seat.
(6) Three doors—one front left; one rear right; one rear left.
(7) Three doors—one front right; one rear right; one rear left.
(8) Three doors—one front right; one front left; one rear right.
(9) Three doors—one front right; one front left; one rear left.
(10) No doors at all—steps provided to get in and out of the body.
(11) Four doors.

From these eleven different combinations there are five that can be considered practical. No. 1, one front right hand and one rear right hand. With a left hand control all passengers and driver can alight on the right side on to the sidewalk; the driver will have to climb over the passenger in the front seat to get out or the passenger will have to get out first unless the front seat is divided. Then the driver can go around and get out of the rear door if divided front seat is provided, No. 3, one front left hand and one rear right hand. In this combination the driver can get out without disturbing anyone, but the front passenger will have to wait

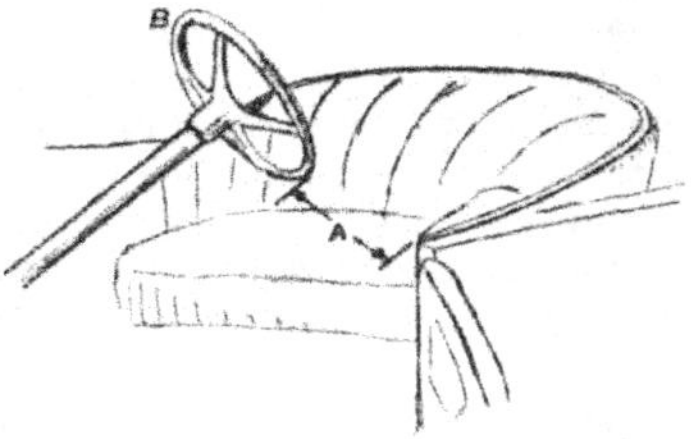

Fig. 128
Relative location of doors, seat and steering wheel

until the driver is out before he can alight, unless there is a divided front seat, which is the better of the two arrangements, if a two-door type of body is to be used. If the front seat is not divided the first arrangement is the better. No. 5 is a divided front seat arrangement, and calls for two rear doors, right and left. No. 7 has one front right and two rear doors right and left, with or without divided front seat. No. 11 combination has four doors.

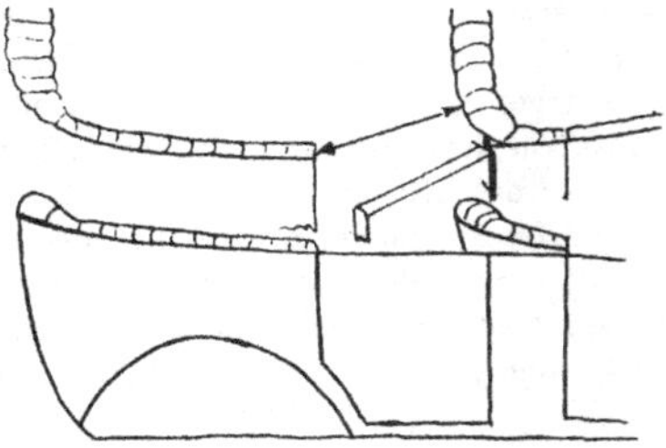

Fig. 129
The door generally extends a little way ahead, as shown here

In the five arrangements enumerated three have two doors, one has three doors and one four doors. The two-door types are used for economical reasons, while the three-door one is used when a tire is carried on the side of the car and blocks the left hand front door.

All of the eleven different methods have been tried at some time or other, but when they have all been carefully considered there is only one satisfactory way and that is to have four doors, because under certain conditions it is found convenient to get out of the car from every side. For instance a left hand control car with a door or doors on the right side will permit the passengers to alight on the sidewalk, but if there is not means for the passengers to alight on the left side there will be a large number of car owners forced to walk around their cars to get into their houses, as a great number of houses have their garages on the left side of the house, and under these conditions he will want to be able to get out on the left side as well as the right. The four-door types should be considered as standard.

The locations of the front doors from the dash are determined by the room required to get into the car without interfering with the steering wheel, and also by the amount of the projection required at the side of the seat. The rear door's position is determined mostly by the distance from the back of the front seat and the position of the fender wheel house. While a wide door is always desired the

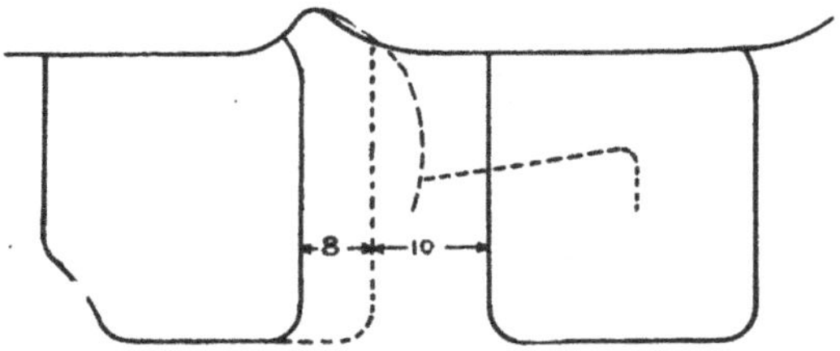

FIG. 130

The double-cowled body cuts down the width of the doors

above factors govern the width of the doors. On a two or three-door car, where the left hand door is omitted, the position of the wheel is not important, but in a four-door car the left hand door is of no use if there is not sufficient clearance between the post and the wheel to allow a person to pass through.

The distance A shown in sketch 128 should be about 12 inches. The dimension B should be sufficient to provide a comfortable side support for the driver. In the rear door the distance C is the actual space available for the passengers to get in and out, but the door generally extends ahead a little way, as shown in Fig. 129. In the average car the door goes back as far as possible to the fender. When there is ample room for a wide door the shape is about the same as the front door, as shown by the dotted lines, but when the wheelbase is short it follows the line of the fender as shown in the full line.

The bottoms of the doors are shaped in two ways, sharp and rounded. The sharp or square door is something that varies with the style. The rounded edge door is used as standard by the majority, as it permits of a stronger door and threshold joints.

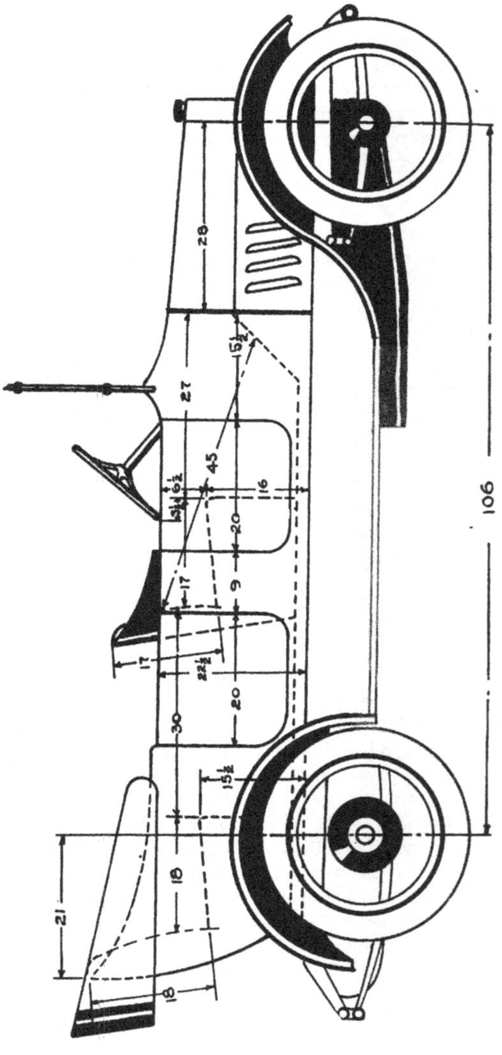

Fig. 131
Small car with dimensions of overhang

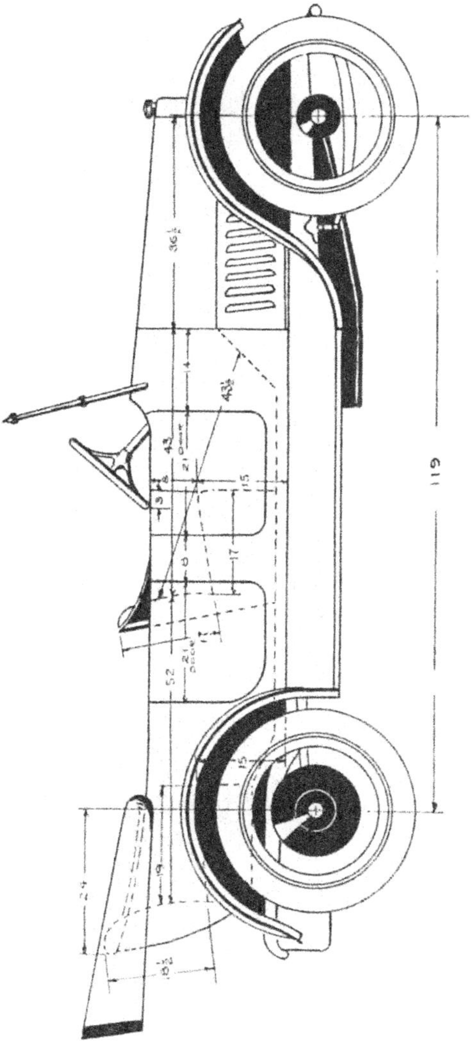

FIG. 132
Small car with dimensions of overhang

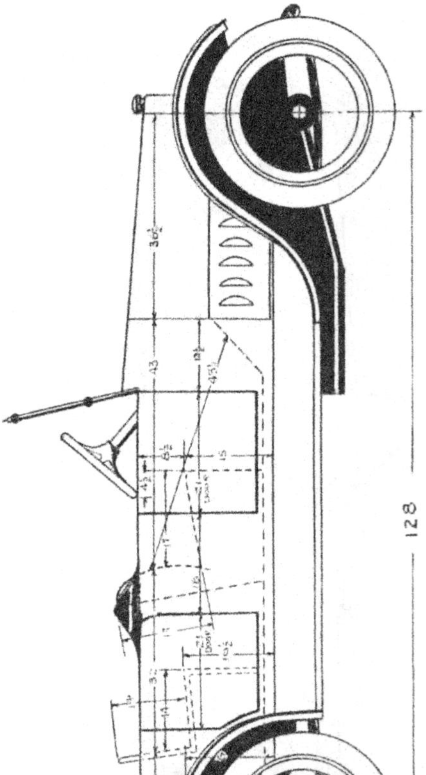

FIG. 133
Small car with dimensions of overhang

The position of the hinges on the doors is another condition that varies. The front and rear doors can either be hinged from the front or rear post; so it is possible to have four different combinations, first front door hinged at front, rear door hinged at front; second, front door hinged at rear, rear door hinged at rear; third, front door hinged at front, rear door hinged at rear; fourth, front door

Fig. 134
Rolls-Royce sport body with fairly high sides

hinged at rear, rear door hinged at front. The most used method is to hang all four doors at the front. The next method, and really the best one, is to hinge the front doors at the front and the rear doors at the rear. Hinging the rear door at the rear gives a much better opening to get through and affords some protection from the rear fenders.

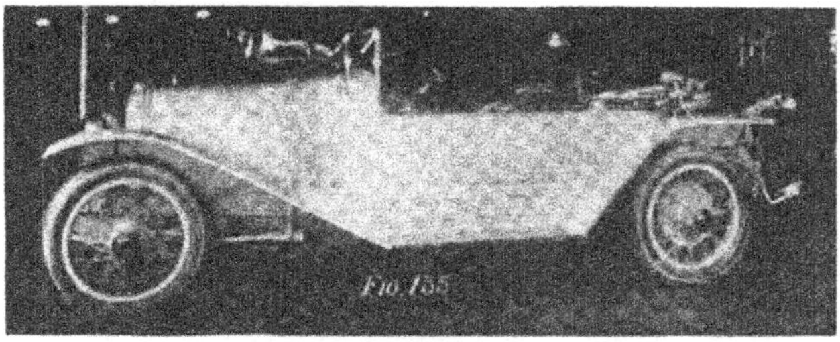

Fig. 135
Low-sided body on Locomobile chassis

The double cowled body cuts down the width of the doors as shown in Fig. 130; the effective opening is reduced by about 8 inches, while the entire door width is reduced about 6 inches. This is shown very clearly by the dotted lines that represent the ordinary seat back, so while the front seat cowl adds greatly to the appearance of the car it reduces the effective door opening at the rear and unless the body has a liberal length this cowl is not practical.

The height of the side of the body is something that has to be determined before the general outline can be laid out. The 1922 bodies have medium high sides and this is possible because the prevailing high radiator and hoods. In sports cars the seats are usually lower so the body sides can be low which tends to give the car a racy or sporty appearance. The overhang of the body as shown in Figs.

Fig. 136
A 1914 Buick model with a well-balanced and gracefully curved body

131, 132, 133, with their dimensions 21, 24 and 23 respectively must be watched closely on a small car. This must be reduced to the limit to save weight but on a large car more overhang can be provided if necessary as it will balance up well.

Fig. 137
A 1922 Buick body with low cut sides

Two good examples of sport bodies are shown in photographs, Figs. 134 and 135. Fig. 134 shows a Rolls-Royce sport body with fairly high sides, while Fig. 135 shows a low sided body on a Locomobile. A 1914 Buick is shown in Fig. 136. This is a well balanced and gracefully curved body. The 1922 Buick body is shown in Fig. 137.

FITTING ENGINE HOOD TO BODY

The method of fitting the engine hood on to the body is an important point and there are three ways how these are generally accomplished. The first method is to depress a ledge in the shroud as shown in Fig. 117, page 51; the hood rests on this depression and leaves a space between it and the shroud for it to work back and forth when the car is in use. The second method is to fit an aluminum rim around the shroud, against which the hood is fitted. When the hood works in service it rubs against the right rim, which does not show at the place where the hood rubs. (See Fig. 118). The third way is to let the hood rest on top of the shroud, which has a lacing attached as shown in Fig. 119.

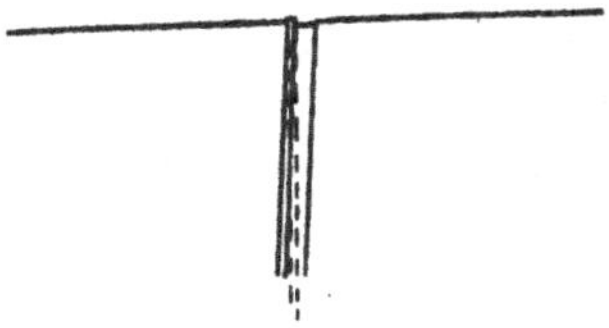

FIG. 138
An example showing hood and shroud out of alignment

The shroud is carefully worked out so that the hood will clear the metal, when it works in service, and not scratch the paint. This last method is the most satisfactory, as it makes a neater appearance than the other two. It also allows for variation in the hood assemblies because there are no lines for the edge of the hood to match, and a slight variation is not noticeable, whereas in the first and second methods, if the hood or the shroud is out of line at all and do not match one another, it is very noticeable, as shown in Fig. 138.

The foregoing covers the features not described in the previous chapters and in order to see what the rules outlined will give, when applied to a touring car body design, a scale drawing of three types of bodies are shown in Figs. 131, 132 and 133. The first is a small body on a short wheelbase of 106 inches, and is typical of the four-cylinder car class. The second is a medium sized car on 119-inch wheelbase, and the third a large seven-passenger six with 128-inch wheelbase. These will cover the average car conditions, and will be good examples of body designs within the limits of their chassis.

ROADSTER BODY DESIGN

The roadster of today has, in most cases, a body that can carry two people and perhaps three, if they are crowded. The roadster has a wide range of utility, its functions running from the business car to the sport or pleasure car. The first bodies used on motor cars were box-like affairs with two seats. After the cars became further developed it was found that 5 to 7 passengers could be transported almost as easily as two, so the bodies were lengthened accordingly. As the automobile was perfected it became more a means for transportation than for pleasure, and this resulted in the 5 and 7-passenger cars becoming the most popular. Consequently the roadster is now almost in the special car class and there is a much smaller number of these made than of the 5 or 7-passenger car. In fact there are some companies that do not make roadsters. The two-seated car is ideal for the doctor or business man who has to do much traveling. It is also ideal for the extremes in sport cars and lends itself to the designer's and builder's skill in planning and making a beautiful car.

The first consideration in a roadster is the two or three-passenger type which is manufactured in regular production. These have the same front ends or cowls as the touring car and in some cases use the same steering wheel and seat position, the only difference being in the rear end and door width. If a lower car is desired than the regular type a larger wheel is used and dropped down. In Fig. 139 is shown a drawing of the average roadster. This is not one of the radical designs, but a type that would be successful in large production, as it will suit a high average of people. The door hinge arrangement can be made in four ways, although there are only two generally used, and these are the first two of the following:

1. Both doors hinged at the front.
2. Both doors hinged at the rear.
3. Right doors hinged at the front and the left at the rear.
4. Left doors hinged at the front and the right at the rear.

Photographs 140 and 141 show the Buick four-cylinder and six-cylinder roadsters. These are typical of the class just mentioned and all show balanced three-

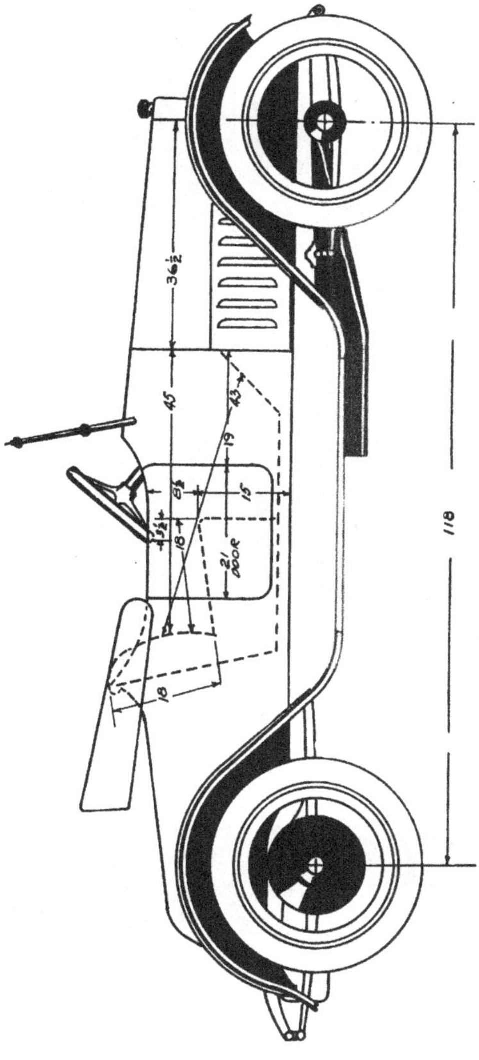

Fig. 139

Type of Roadster that would be successful in large production, as it will suit a high average of people

Fig. 140
Buick Four-cylinder Roadster

Fig. 141
Buick Six-cylinder Roadster

Fig. 142
Large Roadster with offset driver's seat, and room for two passengers on the side seat

passenger single seat roadsters. Two compartments are provided, one inside the body behind the seat back, and the other in the deck. Access is obtained to the latter by a large door in the deck.

In Fig. 142 is shown a large roadster that has an offset driver's seat and room for two passengers on the side seat. This was a very popular type of car and it is a good example of balanced design. There are occasions when there is a desire to carry more than one or two passengers in roadsters so this led to the development of extra or emergency seats. By lengthening the body a third seat can be inserted similar to that roadster shown in plan view in Fig. 143. This type of roadster is called the "clover leaf" because the position of the seats is similar to a three-leaved clover. A good example of clover leaf roadster body design is shown in the photograph 144. This particular car has a compartment in the body for the top to fold into, so the body has a very clean cut appearance

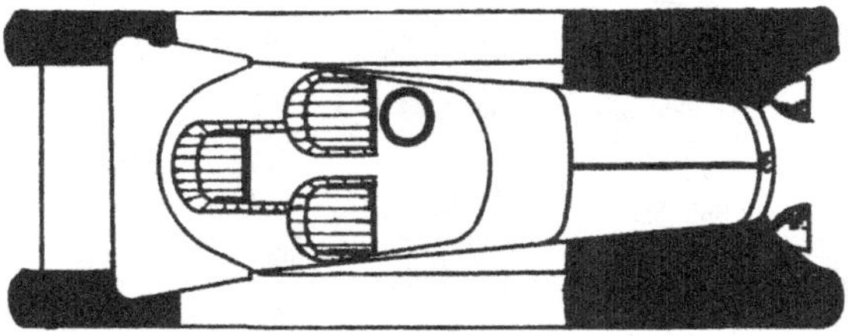

FIG. 143

Type of Roadster called the "clover leaf," because the position of the seats is similar to a three-leaved clover

when the top is down. In order to meet the demand for emergency capacity for the roadster several body designs were worked out so the body could be used for both, a two-passenger roadster and a four or five-passenger car.

Paige Motor Car Co. developed the roadster shown in Fig. 145, 146, 147 and 148. In Fig. 148 the car is shown as a typical two-passenger roadster. The top is concealed so that the body has a smooth and clean appearance. A plan view of the car is shown in Fig. 146, which shows how the rear part is all folded away when it is not in use. When the rear compartment is required it folds out as shown in Fig. 148, providing a very comfortable seat, as shown. When the seat-back is in the position shown there are two doors available, one on each side of the body, so the car is converted into a regular touring car type. The top used with this type of body is unique, as it is really a fabric stretched across bars raised out of the body. This is illustrated in Fig. 148.

Another type of convertible roadster is shown in Figs. 149 and 150. This body has a divided front seat, and when the rear back is lifted up two seats are provided to which access is obtained by going through the divided front seat. The picture, Fig. 149, shows the top up and rear compartment closed; the small

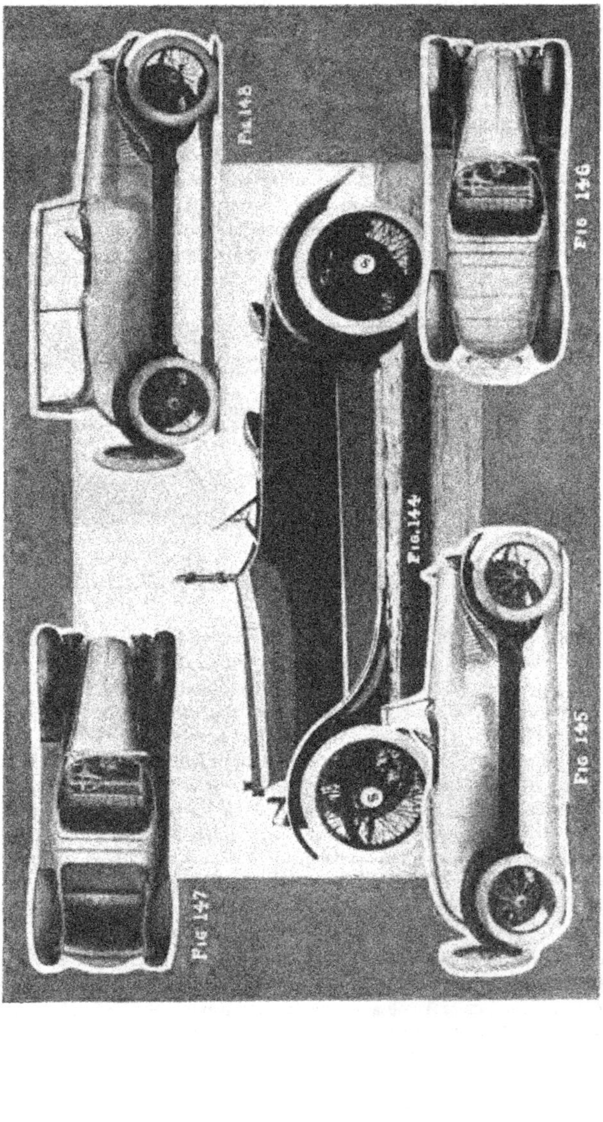

Good example of clover-leaf Roadster body. The Paige Motor Car Company developed the Roadster body shown in Figs. 145, 146, 147 and 148.

picture below shows the car with top drawn and compartment closed. The top is shown up in picture 150, and the seat back raised to provide the two extra seats. The smaller picture below shows the seats in use but the top down. This makes a very convenient arrangement.

Another type of convertible Roadster is shown in Figs. 149 and 150. Fig. 149 shows the top up and rear compartment closed. In Fig. 150 the top is shown up, and the seat back raised to provide two extra seats.

The custom built roadsters frequently have a small seat in the rear compartment as shown in the photograph. Fig. 151. This, of course, is a "fair weather" seat only as the top does not reach over it. Seats are also made to attach to the side of the body and running board. The car displayed in Fig. 151 is a special design and a very sporty appearing car. It will be noted that the running

Fig. 151 has a small seat in the rear compartment while a sharp line Roadster with wide back deck is shown in Figs. 152 and 153. All these are special custom designs and are quite "sporty" in appearance.

boards are left off and aluminum steps provided instead. Such construction is very popular in this class of car.

A sharp line roadster with the wide back deck, which is very popular at this time, is shown in Figs. 152 and 153. This car is a good example of the custom made roadster. The special equipment constitutes one of the selling features over the standard or production roadster. The paint is special selected color and the top is khaki. Cowl lamps, mirror, cigar lighter, special selected radiator cap, plated windshield, cowl ventilators, wide doors, outside door handles and special trimming are the main features to be considered for a special roadster, such as is produced in custom shops.

CHAPTER IX

GENERAL CONSTRUCTION OF MODERN BODIES

The modern body is constructed in a number of different ways. The methods used, and the designs, are governed by the results required. The first type is the combined wood and steel body, and this is the design that is used more than any other and is the one analyzed in this series. The others are as follows: (2) wood and aluminum construction; (3) cast aluminum construction; (4) all steel construction.

The second type is produced much the same way as the steel and wood is, the difference being that the aluminum is used instead of steel, and a heavier wood frame-work has to be made. The main reason for using aluminum is that it is easier to form than steel and when only a few parts or perhaps the body is to be constructed the aluminum can be worked out to a greater advantage. It has the advantage over steel in the fact that it is lighter and if the body is made carefully it can be made lighter than the steel one. The aluminum is rust-proof and is preferred by some on this account.

The third type, the cast aluminum body, is used by a few manufacturers who desire to make a body without regard to expense and produce a design of body that will adapt itself to this construction. The cost of aluminum and the difficulty of casting large panels successfully makes this method of manufacturing bodies prohibitive to the majority of companies.

The fourth type, the all-steel body, is something the manufacturers and designers have been thinking of for a good many years. A number have been made successfully and otherwise; some are being made today and are in use and proving very successful. There are many advantages for an all-steel body. The most important is that it can be enameled and baked. For a low-priced car the operation of enameling is very economical and produces a very fine finish. To enamel a semi-steel body and carry the heat to the proper temperature would prove disastrous to the wood frame.

A few of the problems attached to an all-metal construction are: first, to get the metal formed to obtain uniform sweeps at the various parts. Another condition is to fasten the steel parts together. Spot welding is used for this, but the spots are liable to show on a finished surface and they are also very liable to fail if they are not properly made. As steel is to be substituted for the sills and strainers, etc., the body is liable to be heavier than the semi-steel one.

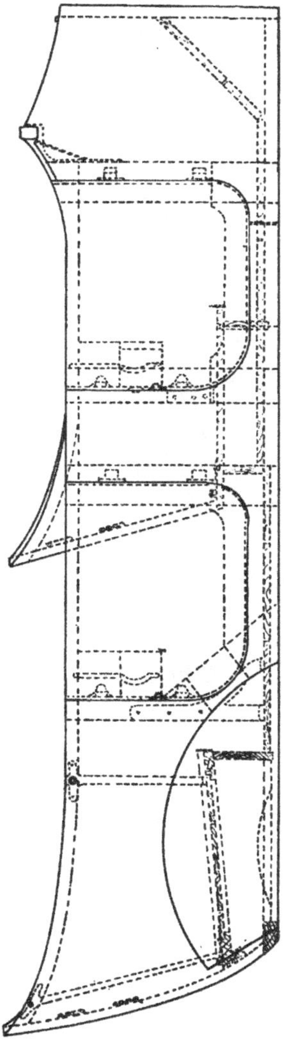

FIG. 154

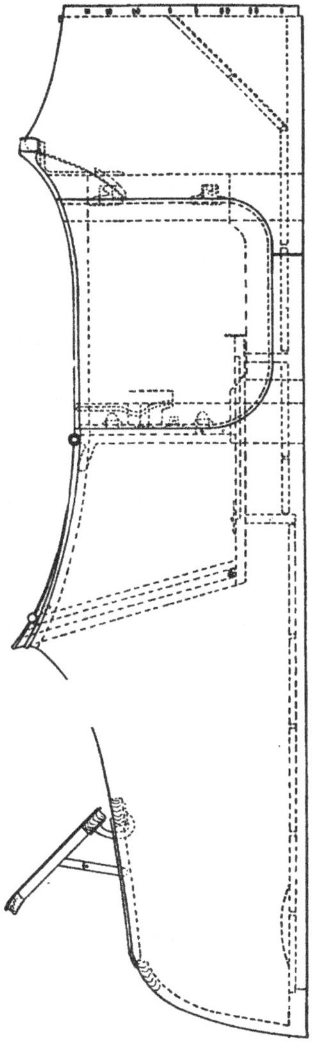

FIG. 155

Fig. 154 and Fig. 155 represent two bodies of semi-steel construction, showing the construction of the bodies and the parts which go to make them complete

The all-steel body is perhaps the body of the future, but up to the present time they have not been designed so that they can be produced any cheaper or more reliable than the semi-steel body. Steel construction, to a certain extent, is very economical, but there is a dividing line where it becomes more expensive to produce an article out of sheet steel than of some other material, such as iron or wood. The price of steel has continued to increase so that it is still the most economical to use wood or other material in preference, as the advantage gained by using steel does not justify the extra expense.

The modern body in its finished state is composed of a great number of parts, so in order to make the study of its construction possible, a system of grouping the units of which it is composed will be outlined. This system will act as a very comprehensive guide. Two bodies of the type of semi-steel construction are to be considered, a touring car and a roadster.

Front end view of body partially disassembled for shipping purposes

Drawings of these are shown in Fig. 154 and 155, these drawings show clearly the construction of the bodies and the parts which go to make it complete. Photographs of a body are shown in Fig. 156 and 157. These show a body partially disassembled for shipping purposes. The unit is very conveniently grouped so that all parts are shown.

The touring car design may be taken up first and divided into groups. Each group will be composed of all the necessary parts that go to make it a unit, so that when all the units are assembled the result is a completely equipped body.

These assemblies or units, of which there are five, are as follows:

1. Shroud Assembly.
2. Side Assemblies.
3. Rear Seat Assemblies.
4. Door Assemblies.
5. Front Seat Assemblies.

In unit No. 1, the shroud assembly, there are the following assemblies and parts:

1.1	Shroud Panel	1.7	Windshield Brackets right
1.2	Dash Panel	1.71	Windshield Brackets left
1.3	Toe Blocks	1.8	Miscellaneous
1.4	Door Post front right	1.81	Hinges right
1.41	Door Post front left	1.811	Hinges left
1.5	Shroud Bar	1.9	Toe Boards
1.6	Shroud Bar Bracket	1.91	Floor Boards

In unit No. 2, side body assemblies, there is a right and a left group. These will be listed together as follows.

2.1	Side panel right	2.351	Threshold bar rear left
2.11	Side panel left	1.8	Miscellaneous
2.2	Sill right	1.81	Hinges right
2.21	Sill left	1.811	Hinges left
2.3	Door post front rear right	1.82	Lock strikes
2.31	Door post front rear left	1.83	Bumpers
2.32	Door post rear front right	1.84	Dove tails female
2.321	Door post rear front left	2.40	Door jambs
2.33	Door post bar front right	2.41	Door plates
2.331	Door post bar front left	2.5	Scuff plates front right
2.34	Threshold bar front right	2.51	Scuff plates front left
2.341	Threshold bar front left	2.6	Scuff plates rear right
2.35	Threshold bar rear right	2.61	Scuff plates rear left

In unit No. 3, rear seat assemblies, there are the following assemblies and parts:

3.1	Rear seat panel	3.54	Seat bottom
3.2	Door post rear, rear right	3.55	Heal board or filler
3.21	Door post rear, rear left	3.6	Floor boards
3.3	Door post brace right	1.8	Miscellaneous
3.31	Door post brace left	1.82	Lock striker
3.4	Top rail	1.88	Top iron right
3.41	Trim rail	1.881	Top iron left
3.42	Cross end sill	1.84	Dove tail female
3.43	Strainers and trimming spring slats	1.83	Bumpers
		1.87	Hinges seat lid
3.5	Seat frame	1.89	Top holder iron right
3.51	Seat raiser front	1.891	Top holder iron left
3.52	Seat raiser rear	1.9	Seat trimming blocks
3.53	Seat lid	1.91	Door jambs

In unit No. 4 door, there are four assemblies:

4.1 Front door right
4.11 Front door left
4.2 Rear door right
4.21 Rear door left
4.1 Doors front right and left
4.3 Front door panel right
4.31 Front door panel left
4.4 Front door frame right
4.41 Front door frame left
1.8 Miscellaneous
1.85 Door dove tail male
1.86 Lock right
1.861 Lock left
4.2 Rear door right
4.21 Rear door left
4.5 Rear door panel right
4.51 Rear door panel left
4.6 Rear door frame right
4.61 Rear door frame left
1.8 Miscellaneous
1.85 Door dove tail male
1.86 Lock right
1.861 Lock left

Side view of body partially disassembled for shipping purposes

In unit No. 5, front seat assemblies, there are the following parts and assemblies:

5.1 Front seat panel
5.2 Top rail
5.21 Trim rail
5.22 Strainer and trimming spring slats
5.23 Seat risers front
5.231 Seat risers rear
5.3 Seat frame
5.31 Seat lid
5.32 Seat bottom boards
5.4 Door post brace
1.8 Miscellaneous
1.87 Hinge seat lid
1.9 Seat trimming blocks

Having described the various groups and parts that go to make them up, there is another grouping that can be made, and that is of the materials used. For instance, the body is composed of wood, sheet steel, malleable iron, or steel castings and steel forgings, and machined steel parts. These can be conveniently grouped as follows:

No. 1 Body frame wood
No. 2 Body sheet steel parts
No. 3 Body castings and forgings
No. 4 Miscellaneous parts classified as hardware, such as locks, hinges, strikes, dove tails, bumpers, etc.

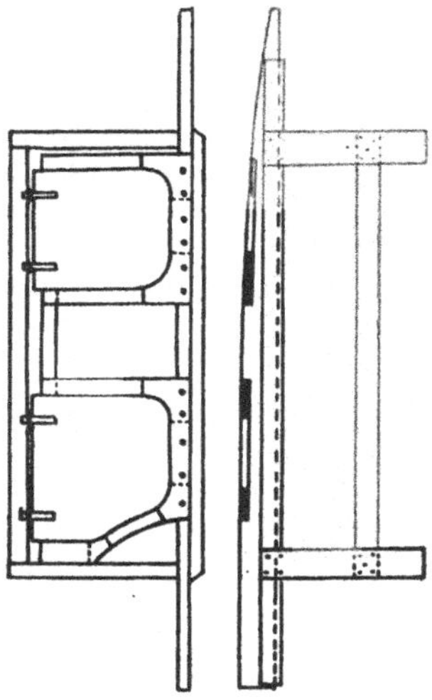

FIG. 158

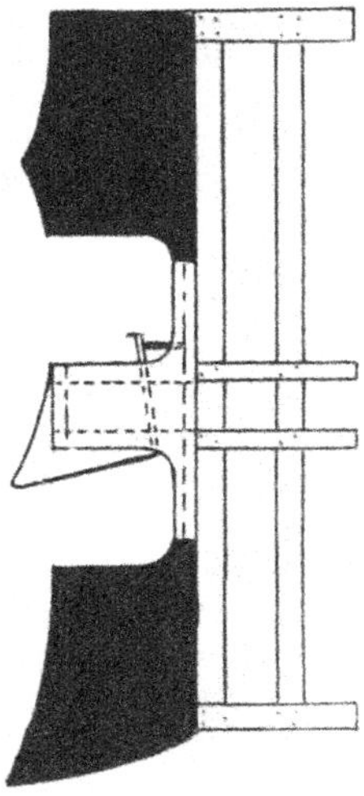

FIG. 160

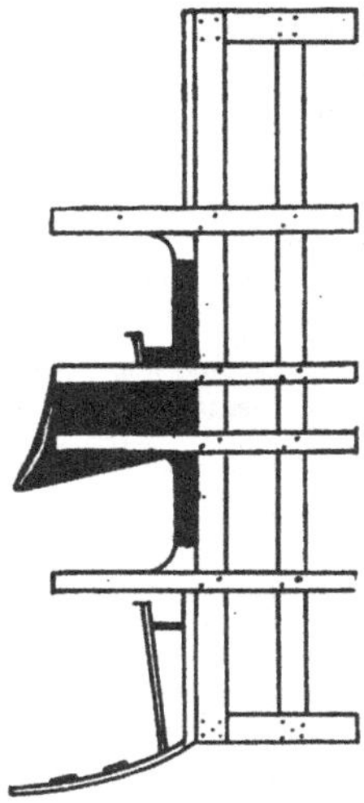

FIG. 159

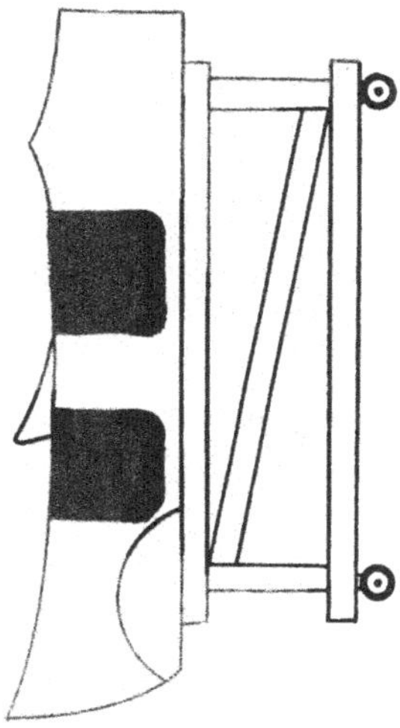

FIG. 161

Figs. 158 to 161 show the forms or jigs used in body assembly. An explanation of the purpose of each of these forms is given on page 80

Groups 1 and 2 will be treated in detail in a subsequent chapter and also groups 3 and 4 as there is a great deal of work on these parts to make them complete for assembling, and the details of their manufacture is very interesting. The method of enumerating these groups is the decimal system and it will make a convenient way of classifying details and data on these various parts. In Figs. 154 and 155 the parts and their groups are shown, giving a clear idea of the relations of the different parts being described. It should be remarked, however, that in manufacturing bodies the parts will not all be able to be assembled in the factory in the way enumerated, as different problems come up that call for special arrangement of parts. This grouping was composed for reference, parts list and specification work, more than for manufacturing assembling. The method of assembling the above body can be outlined as follows:

The side sills are set into a jig, see Fig. 158, and the door posts, door post bars, side panels and thresholds are assembled to them. At the same time the hinge sections, dove tails and strikes are also assembled to the posts, a gauge being used so that they are accurately located. These sill assemblies are next set into another form like that shown in Fig. 159, and clamped into position so that another stage of assembling can be accomplished. In this second stage the shroud bar and bracket, the shroud panel, windshield bracket, and the front seat assembly are attached. The body is next moved to the third form where the rear seat frame, strainers and back panel are fitted as shown in Fig. 160.

The fourth form used is for assembling the front and rear top irons and when this is done the body is ready for the door hanging operation which can be termed the sixth operation. The doors go through a separate operation of assembling which is of interest, and that is fitting the panels, lock, hinges, dove tails and strikes. A suitable clamping arrangement is provided to hold the panel in place while the door is being nailed, and while the door is still clamped in position the hinges and the other parts are located accurately by means of suitable gauges. To hang the doors, as this fitting operation is called, the body is set on a truck as shown in Fig. 161, and this same truck carries the body through the seventh and final operation of finishing, which is the operation of filing and soldering up all the bumps and dents.

The foregoing covers the parts used and the methods of construction of a touring car body, the roadster will be subject to the same conditions and will use identical parts to a certain extent. In order to have a list to check the roadster parts against a typical roadster, body construction will be outlined in the following six headings:

1. Shroud assembly
2. Side assemblies
3. Rear seat assembly
4. Door assemblies
5. Front seat assembly
6. Rear deck assembly

The unit No. 1 shroud assembly is composed of the same parts as listed on the touring car. The unit No. 2 will be composed of the following:

2.1	Side panel right	2.4	Door jambs
2.11	Side panel left	2.5	Scuff plate right
2.2	Sill right	2.51	Scuff plate left
2.21	Sill left	1.8	Miscellaneous
2.3	Door post front rear right	1.82	Lock strike
2.31	Door post front rear left	1.83	Bumpers
2.34	Threshold bar right	1.85	Dove tail female
2.341	Threshold bar left		

In unit No. 3 there are no parts classified in this design; if it were a chummy or clover leaf design there would be a grouping under No. 3.

In unit No. 4 there are the following parts:

4.1	Front door right complete	4.41	Front door frame left
4.11	Front door left complete	1.8	Miscellaneous
4.3	Front door panel right	1.85	Door dove tails male
4.31	Front door panel left	1.86	Lock right
4.4	Front door frame right	1.861	Lock left

The unit No. 5 front seat is slightly different from that of a touring car. The front seat is not used, so the list starts with:

5.2	Top rail	5.4	Door post brace
5.21	Top trim rail	1.8	Miscellaneous
5.22	Strainer and trimming spring cleats	1.87	Hinge seat lid
		1.88	Top iron right
5.23	Seat riser front	1.881	Top iron left
5.24	Seat riser rear	1.89	Top holder iron right
5.3	Seat frame	1.891	Top holder iron left
5.31	Seat lid	1.9	Seat trimming blocks
5.32	Seat bottom boards		

Unit No. 6 is an additional assembly and composes:

6.1	Deck panel	1.8	Miscellaneous
6.2	Deck panel frame	1.83	Bumpers
6.3	Deck door complete	1.91	Hinge deck door
6.31	Deck door panel	1.92	Lock deck door
6.32	Deck door frame	1.93	Deck door prop

This completes the list for the roadster shown in Fig. 155. The construction which has been outlined in these lists is taken from a modern type of body that is produced in great numbers. The majority of bodies produced in America are made very similar to the construction outlined, although every designer will have detail modification to suit the conditions and his ideas.

CHAPTER X

BODY WOOD WORK

The first automobile bodies were constructed entirely of wood, but it was soon found that metal panels could be formed up much easier than wood ones and they replaced the outer wood panels for general body construction. With the use of metal panels, the wood construction of a body came to be confined to a frame upon which the metal was attached. This is the principle upon which the majority of bodies are constructed today. In the previous chapter the body was divided into five units. These units had the following wood parts in them.

Unit No. 1—Shroud assembly

1.4 Door post front right
1.41 Door post front left
1.5 Shroud bar
1.9 Toe boards
1.91 Floor boards

Unit No. 2—Side body assembly

2.2 Sill right
2.21 Sill left
2.3 Door post front rear right
2.31 Door post front rear left
2.32 Door post rear front right
2.321 Door post rear front left
2.33 Door post bar front right
2.331 Door post bar front left
2.34 Threshold bar front right
2.341 Threshold bar front left
2.35 Threshold bar rear right
2.351 Threshold bar rear left
1.91 Door jambs

Unit No. 4—Doors

4.4 Front door frame right
4.41 Front door frame left
4.6 Rear door frame right
4.61 Rear door frame left

Unit No. 5—Front seat assembly

5.2 Top rail
5.21 Trim rail
5.22 Strainers and trimming spring slats
5.23 Seat raiser front
5.231 Seat raiser rear
5.3 Seat frame
5.31 Seat lid
5.32 Seat bottom boards
1.8 Miscellaneous
1.9 Seat trimming blocks

By outlining the wood parts in this manner the various pieces can be taken in proper order and be described and analyzed.

In unit No. 1 the first part to be considered will be the shroud bar, shown in Fig. 162. This can be made from ash, elm or maple wood and is a plain bar shaped to suit the shroud under which it sets. The toe and floor boards are made from maple wood and fit in the rabbet provided in the toe blocks and the front end of the sill. They are generally made in four sections, called the upper and

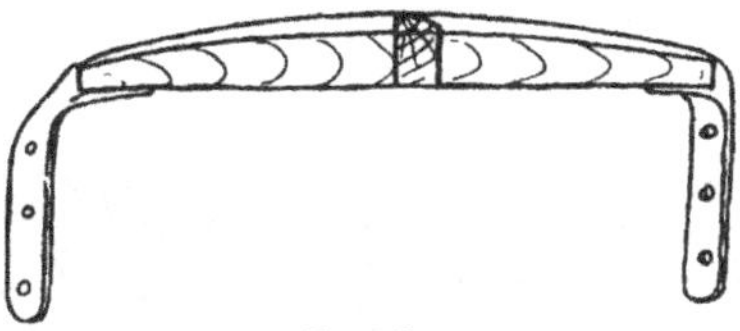

FIG. 162
Shroud bar

lower toe boards and the front and rear floor boards. The toe boards have to have cut-outs in them to clear the various parts of the car that project through them, such as the steering column, foot pedals, starting pedals, etc. The floor boards may have to have cut-outs in them for the hand control and brake levers,

FIG. 163
Typical set of floor boards

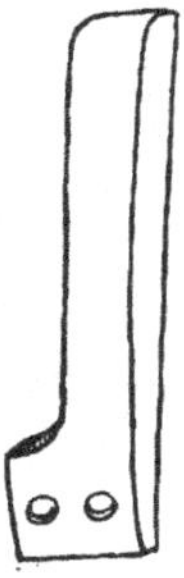

FIG. 164
Door post made by the "separate piece" method

also muffler cut-out pedal. A typical set of floor boards is shown in Fig. 163. The front door post is made of ash, elm or maple, and is shaped on the outside to give the desired sweep to the body, and on the inside for the correct flare for the door.

There are two ways of making these door posts. The first is to make them in separate pieces as shown in Fig. 164, and the other method is to machine the door post door frame and threshold out of one frame. The door is band sawn out of the frame. The advantage of this method is obvious as a great deal of time, labor and material is saved. The older method would entail wasting stock on each post machined and then the door would have to be framed and machined separately. This line-up made a considerable number of operations.

The new method of machining developed, perhaps, from the machines made to shape door frames. This machine consists of two knives made to revolve at a very high speed; the knives are so made that they will cut the desired sweep

The door frame is securely clamped in a heavy frame that slides on irregularly shaped guides that give the desired sweeps while the wood is being machined.

on the door frame as it passes over them. The door frame is securely clamped in a heavy frame that slides on irregularly shaped guides that give the desired twist in the sweeps, while the wood is being machined. This is shown in Fig. 165. This method of machining was perfected to the extent that it is possible to machine the door and the door post all at one time. This, of course, made the machine larger, but now one pass over the knives completes door post and door together. The frame-up is made like that shown in Figs. 166 and 167.

Fig. 167 shows the door post which remains after the door frame, which is shown in Fig. 166, is cut out. The completed door is shown in Figs. 168 and 169. The door is shaped on the outer edges in a form as shown in Fig. 170, and in Fig. 171 it is shown being hand sawn out of the frame.

In describing the methods of door making all the door posts, door frames, and thresholds are fully covered, so that in taking unit No. 2 the only parts to need descriptions are the sills, post bars and door jambs. The sills are the main part of the body frame as all the body structure rests upon them. The chassis

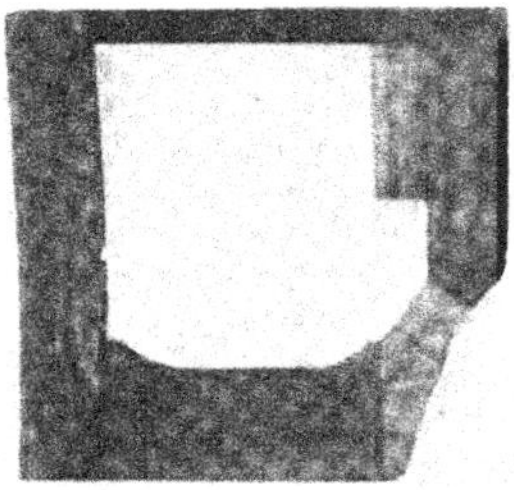

FIG. 166

FIG 169

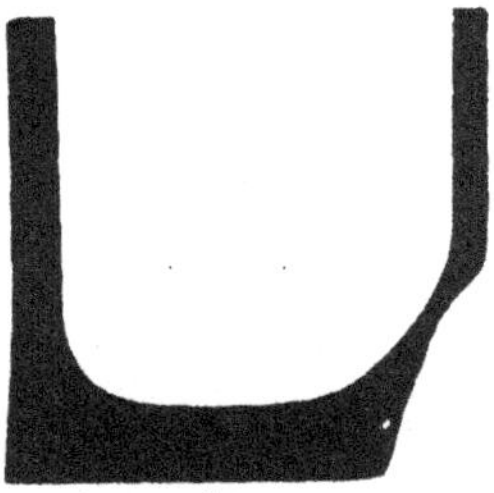

FIG 167

FIG 168

These four figures show the frame-up of a door, Figs. 168 and 169 picturing the completed frame.

frames vary a good deal, so that means various types of sills. The two most common are an off-set or up-set sill and a plain flat sill. The flat sill as shown in Fig. 172 is simply constructed from one piece of timber, machined to suit the conditions required, and is generally made two inches thick. The up-set sill has to be built up from three pieces as shown in Fig. 173.

There are some sills constructed that are built up like that shown in Fig. 174. This constitutes a very strong sill but makes it both thicker and heavier than the solid construction of about 1⅝-inch thickness.

In unit No. 3 there is the top rail and trim rail cross end, sill strainers and trimming spring slats, seat frame and raisers. The top rail is made from ash

The door is shaped on its outer edges in a form

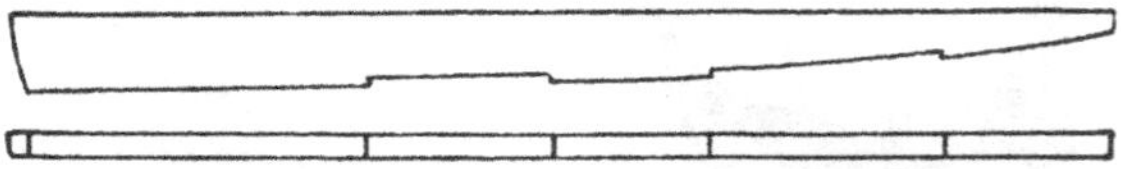

FIG. 172

FIG. 173

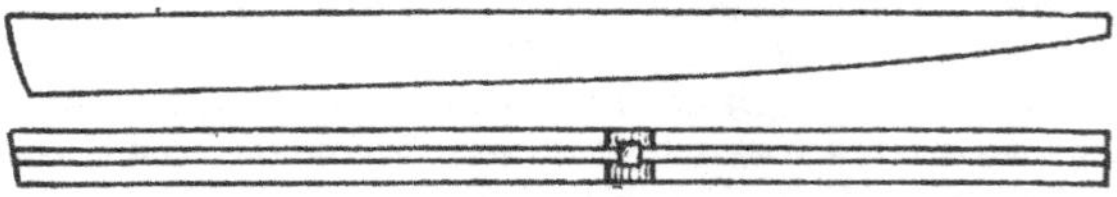

FIG. 174

Various forms of Sills. Fig. 172, Flat Sill; Fig. 173, Upset Sill; Fig. 174, Built-up Sill bent up to the desired shape; the trimming rail is bent, too, in this case, as it comes round the corners. The cross end sill is of maple or ash and is a straight piece that requires no comments, as it is a simple machine job. The strainers are maple or ash, machined to conform with the sweep of the back panel. The slats are straight pieces attached to the strainers in a convenient position for the trim-

One of the processes of door-making. Sawing the door out of the frame

ming springs. The seat frame is a plain frame of ash or maple, machined to the desired shape.

The unit No. 4 has been covered by the door description and calls for no further comments.

In unit No. 5 the conditions are similar to those described in the third unit, so that further description is unnecessary.

CHAPTER XI

BODY METAL WORK

In dealing with the metal parts of the bodies the sheet steel pieces will be taken first and then the forgings and the castings. In the previous chapter the various ways of making a body were taken up and analyzed, and in order to avoid duplication in the detail descriptions of all the different designs of the parts one acceptable design will be treated only. In doing this the details and their construction can be described in proper rotation. The type of body under consideration is the one shown in Fig. 154.

The sheet steel parts used in this body are shown in the following list:

Shroud
Dash
Hood ledge
Instrumental plate
Toe block
Front seat heel board
Front seat corner panel
Front seat back panel
Side panel right
Side panel left
Door panel front right
Door panel front left
Door panel rear right
Door panel rear left
Rear seat side panel, right and left
Rear seat back panel
Rear seat bottom plate
Rear seat heel board and raiser
Cushion retainer front
Cushion retainer rear
Scuff plate front right
Scuff plate front left
Scuff plate rear right
Scuff plate rear left

The shroud is one of the most difficult pieces to make of all the metal work on the body. It generally has a very complex surface which taxes the die maker's skill in reproducing. The Fig. 175 shows the shape of a typical shroud in the condition that it is after coming from the dies, and before the final minor operations are made on it. This is the way the shroud looks coming from the die that forms it in one piece.

To produce shrouds in this manner a huge press is necessary, and very costly dies, and unless the quantity to be made is very large they cannot be produced economically this way. Fig. 176 shows a shroud die in the press. The methods used for smaller numbers is to make the shroud in three sections, viz.: First, the top; second and third, the right and left sides. These pieces are welded together by the oxyacetylene process, and the joints hammered down and filed smooth. In

Fig. 177 the sections, where the joints are made, are shown by heavy lines. A welding form is made to set the three pieces into, so that they are accurately located before being welded.

These forms can be made in two ways. The first, and perhaps the simplest, when small quantities are to be made, is like that shown in Fig. 178. This consists of a wood form shaped to fit the inside of the metal, and at the point where the joints are made iron castings are let into the wood. The hinged strap-like casting is swung down and clamped over the metal so that it securely holds it on each side of the joint. Then the welding torch is applied along the gap in the center where the two ends of the metal meet. When the welding is made these straps are swung out of the way and the shroud is ready for the next operation of hammering and filing.

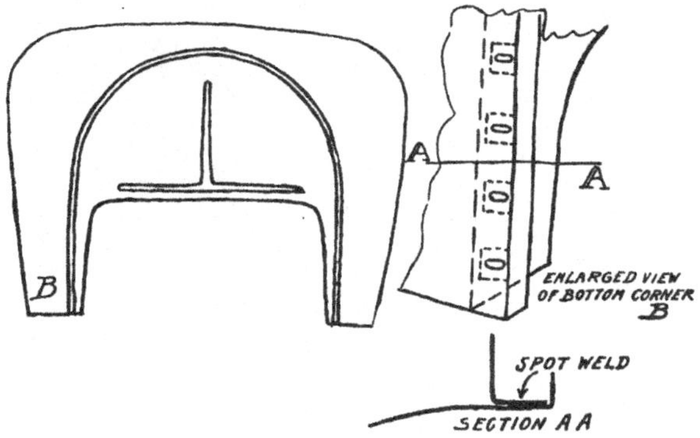

Fig. 175
Shape of a typical shroud in the condition in which it comes off the dies and before final operations are made on it

The dash is next to be attached and this is accomplished by spot welding the flange to the shroud as shown in Fig. 175 and sections. After this is done the slots for the hood ledge lace are punched in and then the shroud is ready for assembling to the body. On this particular design of shroud the flange on the rear at the top of the hinge post and the flange at the bottom of the sill have to be turned over by hand when it is being attached to the body. The dash in some cases is made from one piece of metal and in others two pieces are used. In section Fig. 179 is shown the construction of a one-piece dash, and in section Fig. 180, the two-piece construction. In the first instance the flange, by which it is attached to the cowl, is formed up all in one piece; with the dash in the second instance it is a separate piece, spot welded or riveted to the dash. The difference between these two methods is that the two-piece construction does not take as large a die as the other and is more economical for small quantities.

The holes and cut-outs have to be made to suit the instruments, etc., as required in the design. The large opening is to provide clearance for the steering column and foot pedals. It will be noted that a flange is provided all around this large cutout and that ribs are pressed at convenient places in order to stiffen up the metal and permit a light gauge being used. No. 16 USS gauge is usually satisfactory for this part.

The instrument plate in this design is made of No. 18 USS gauge and is formed to fit the body shroud bar, and the ends are flanged to fit on the face of the

FIG. 176

The blank for a shroud is here shown in position in the die ready for the punch to come down.

door post. The screws and bolts that attach the hinge to the body go through the flanges. The cutouts are provided as required by the instruments. Where steering column brackets attach, a 3/32 plate is spot-welded so that ample reinforcement is provided for the bolts that fasten the bracket to the plate. If this reinforcement were not used the bolts would tear an elongated hole in the light metal and the steering column would work loose. The construction of this plate is clearly shown in Fig. 181.

Toe Block. These can be constructed like the design shown in Fig. 182. No. 18 USS gauge steel is used. The outer edge is shaped to conform with the shroud; the inside is depressed to form a combination of the rabbet and the sills

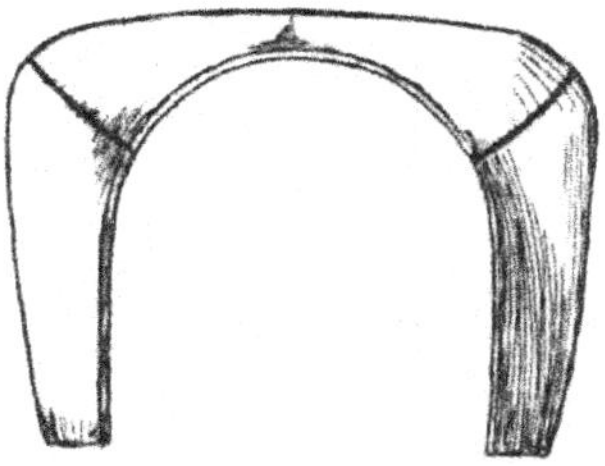

FIG. 177
Shroud made in three sections, welded together

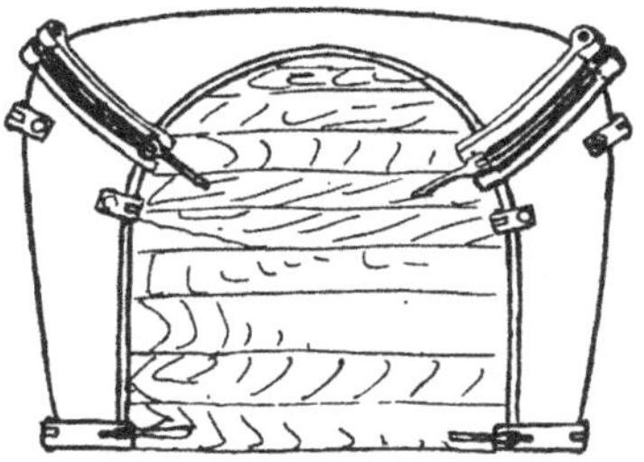

FIG. 178
Welding form used in welding three-piece shroud shown in Fig. 177

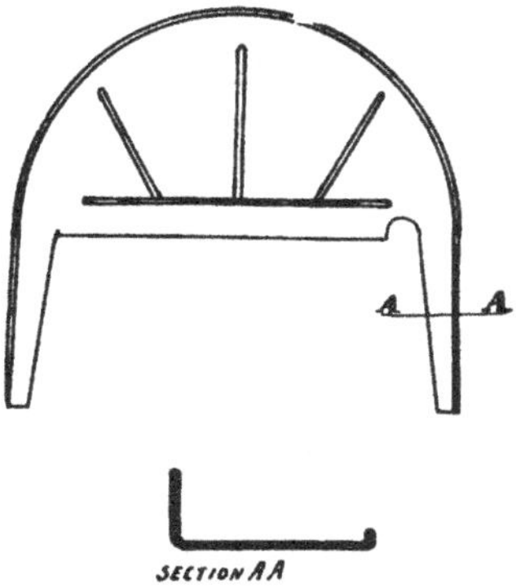

FIG. 179
Construction of a one piece dash

FIG. 180
Two-piece dash construction

in which the toe and floor boards rest. These blocks are riveted to the dash at the top and attached to the sills by wood screws.

Front seat heel board. This heel board or seat riser can be made of all steel or may have steel corner brackets only. The first construction is shown in Fig.

183 and the second in Fig. 184. The first method is slightly stronger than the all steel board, as the wood board forms a tie brace for the body sills. The sketches shown on next page are self-explanatory.

Front Seat Back. This is a simply constructed panel with or without a flange on the top, according to the design. In Fig. 185 is shown a panel with a top flange

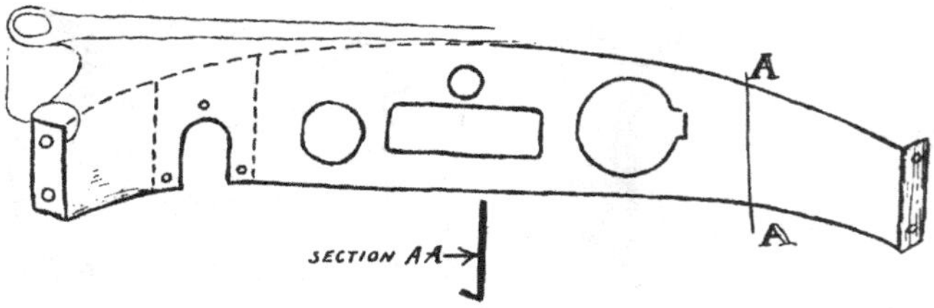

FIG. 181
Construction of reinforcement plate for steering column bracket

having a decided "round" on the corners, and in Fig. 186 is shown outline of a panel that is a flat plate. In the first design the panel is painted, so that the flange has to be nicely finished, but in the second design the back is covered entirely with leather and an unfinished plate is sufficient.

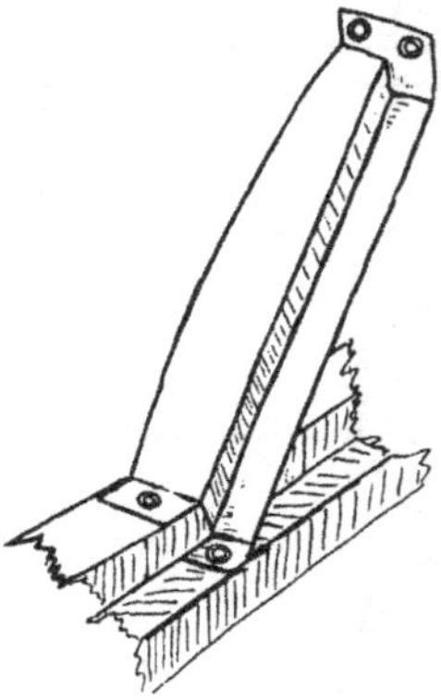

FIG. 182
Toe block, riveted to the dash at the top and attached to the sills by wood screws

Side Panels. These are formed in the dies to fit on the front door post and rear door post, and have flanges drawn on them on four sides. See Fig. 187. This shows a typical side panel. This is a difficult stamping to make as the flanges must all be accurately formed and the sweep of the body has to be properly matched up. The radii in the corners must be properly formed; otherwise it is certain the doors will not fit well.

Door Panels. These consist of plates that are formed to conform with the sweeps of the body and are from ¼ to 7/16-inch larger than the openings in the

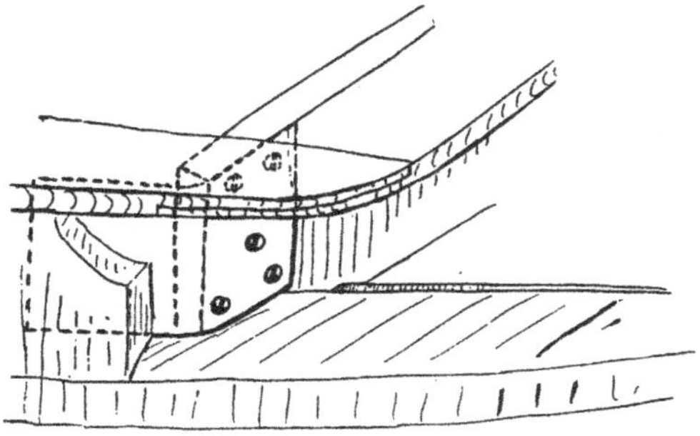

FIG. 183
Front seat heel board. First method of construction

bodies for the doors. They attach to the door frames by suitable flanges. There are two methods of making these, in general use; the first and older method is specially adaptable to small quantities and its construction made is shown in the

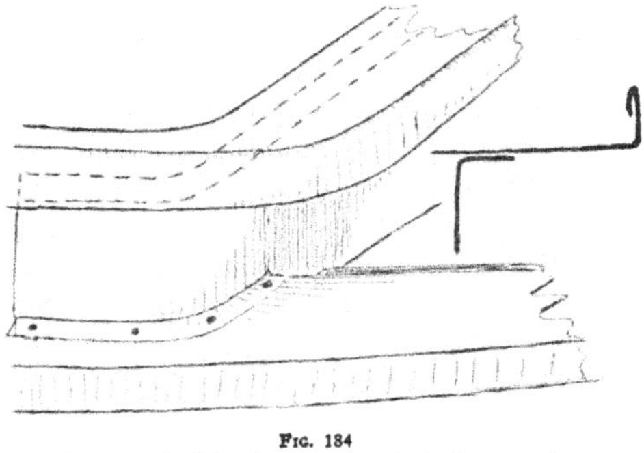

FIG. 184
Front seat heel board. Second method of construction

section Fig. 188. The outer plate is bent down on to an angle and the angle provides the means of attaching it to the door frame.

In Fig. 188 is shown the method of making the door panel and flange out of one piece of steel. This is the method perfected and patented and known as the England door panel. The process of making panels this way is very economical for large production only, as the die equipment is expensive. All the four doors, front right and left, and rear right and left, are constructed and made the same way.

Rear Seat Panel. This is the largest of all metal panels and requires a lot of careful die work to make it properly. The general way to construct this is to

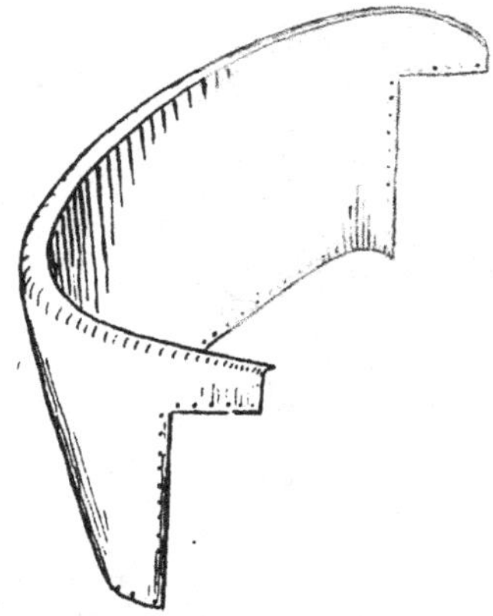

Fig. 185
Panel with a top flange, having a decided "round" on the corners

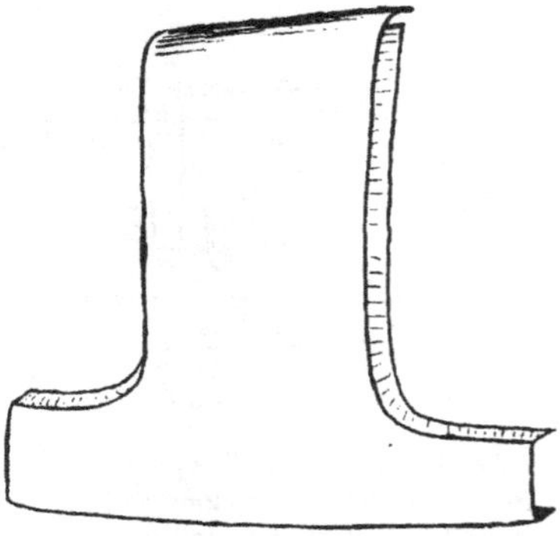

Fig. 187
Side panel formed on the dies to fit on the front door post and rear door post, having flanges drawn on four sides

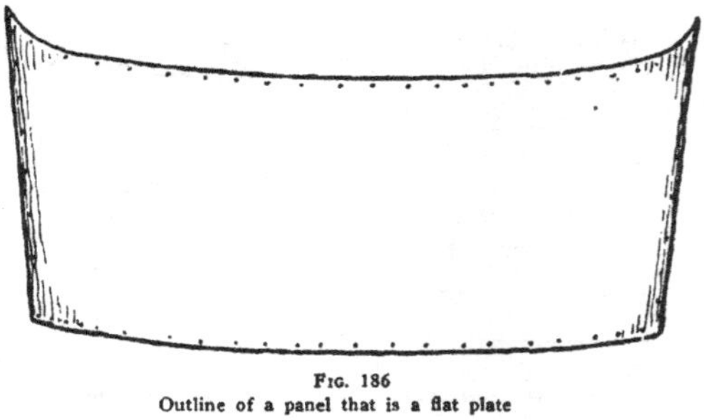

Fig. 186
Outline of a panel that is a flat plate

make it of three pieces, two sides and one rear piece, as shown in Fig. 189. This requires three sets of dies and fixtures. These three pieces are welded together by oxy-acetylene torch in a similar manner to the method described for making a shroud in three pieces. A welding form is made as shown in Fig. 190, and into this the panels are located by straps so that the assembly will be accurate. Clamping straps are swung down and hold the metal firmly while the two sections are being welded together. This joint has to be hammered down and filed smooth.

There is another way to make this panel and that is to form the side and back out of one piece and insert the depression for the wheels. This makes a three-

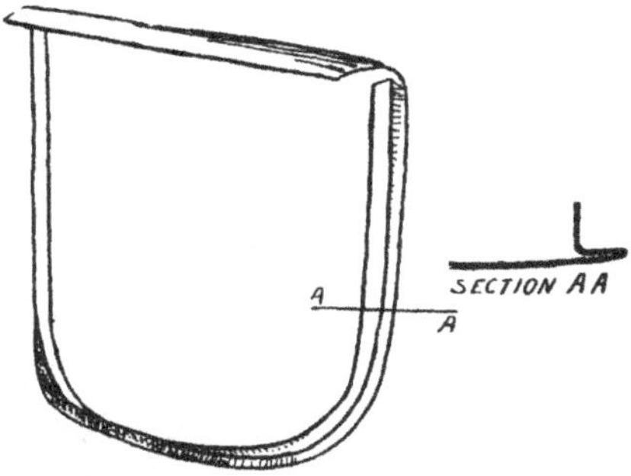

FIG. 188
The outer plate is bent down on to an angle and the angle provides the means of attachment to the door frame

piece back, but it eliminates the welded joint in the back. These wheelhouse sections are spot-welded in position, and this is a simple operation. This construction is shown in Fig. 191. The drawback to this method is the difficulty of getting a

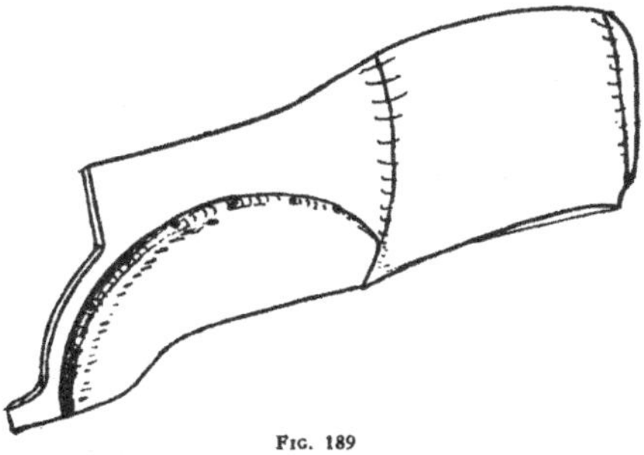

FIG. 189
Rear panel made in three pieces, two sides and one rear piece

properly shaped panel. Some manufacturers are producing backs successfully this way.

Rear Seat Bottom. This is a flat panel made from No. 18 USS gauge fitted under the rear seat to provide a compartment for carrying things. The principal reason for making this of steel is that it gives a little more space than if wood were used and it permits making a depression in it for the axle housing, which will rise up under extreme deflections of the springs and require additional clearance. This seat bottom is shown in Fig. 192. The hump will be seen and also the radiating ribs. These are pressed in for a double purpose; the first is to stiffen the panel and the other is to prevent rumbling when the body "weaves."

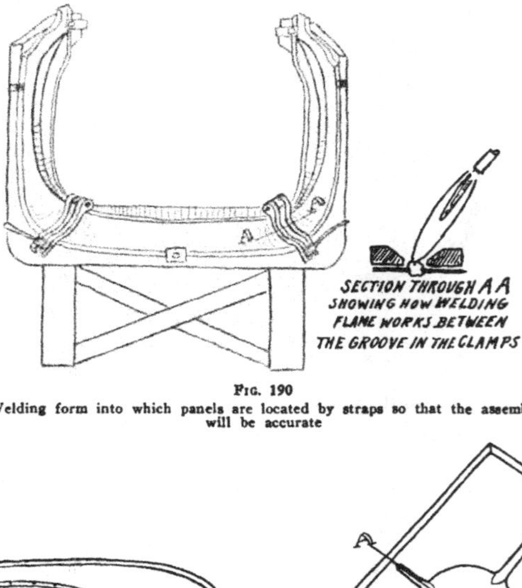

Fig. 190
Welding form into which panels are located by straps so that the assembly will be accurate

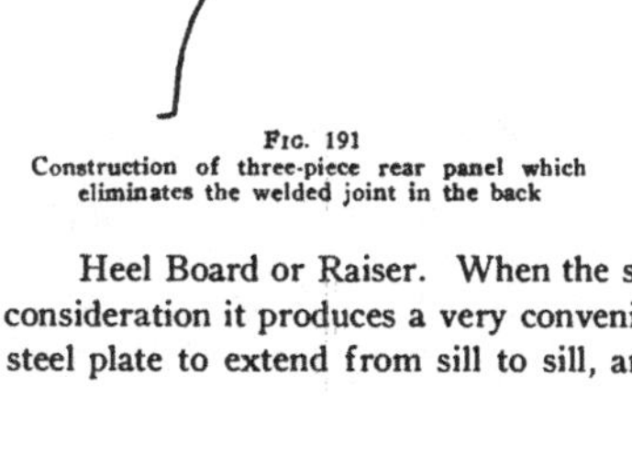

Fig. 191
Construction of three-piece rear panel which eliminates the welded joint in the back

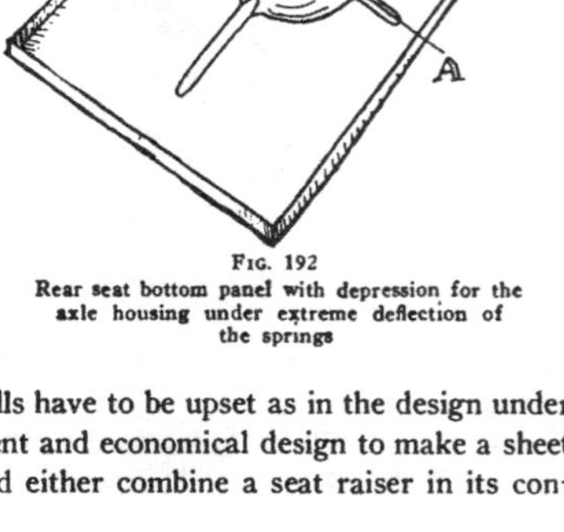

Fig. 192
Rear seat bottom panel with depression for the axle housing under extreme deflection of the springs

Heel Board or Raiser. When the sills have to be upset as in the design under consideration it produces a very convenient and economical design to make a sheet steel plate to extend from sill to sill, and either combine a seat raiser in its con-

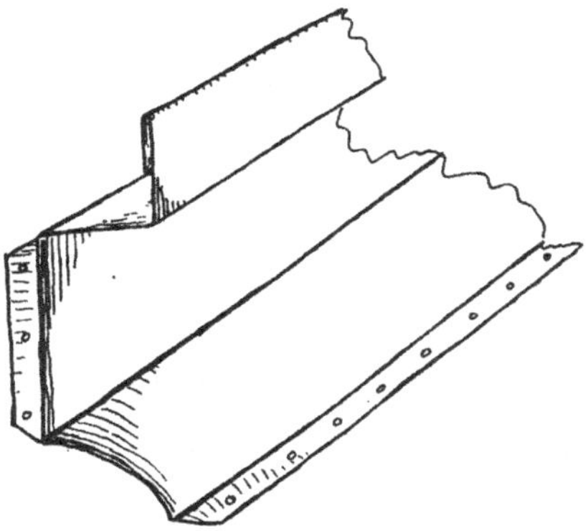

Fig. 193
Typical design of heel board or raiser

Fig. 194
Cushion retainer

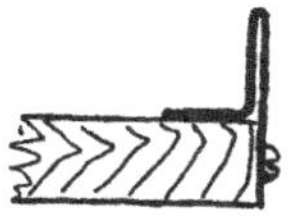

Fig. 195
Another design for cushion retainer

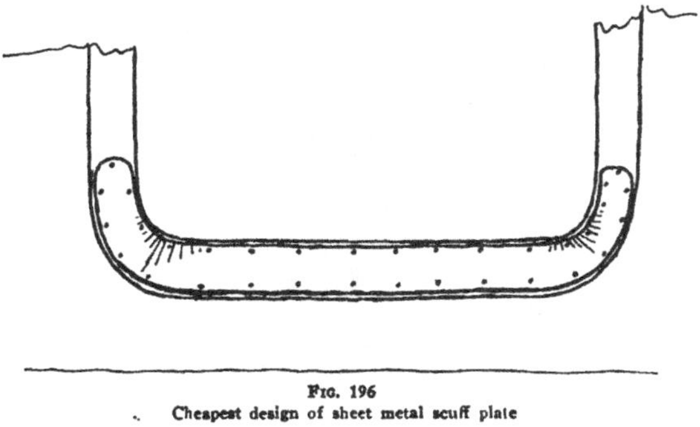

Fig. 196
Cheapest design of sheet metal scuff plate

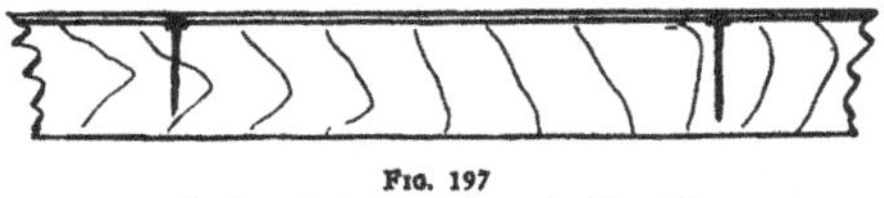

Fig. 197
Section of drawing shown in Fig. 196

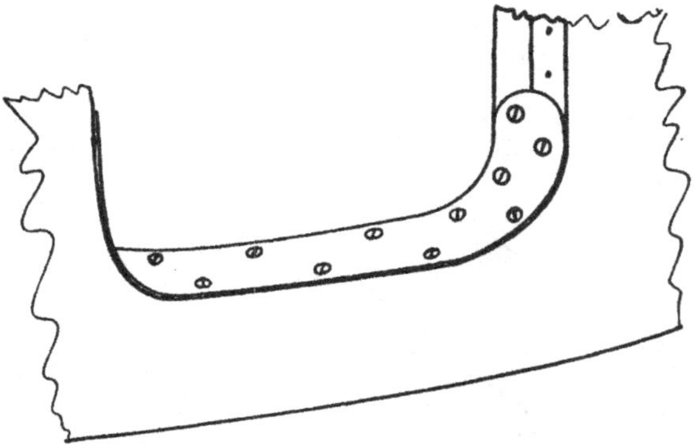

Fig. 198
Scuff plate of polished aluminum, attached by French head wood screws

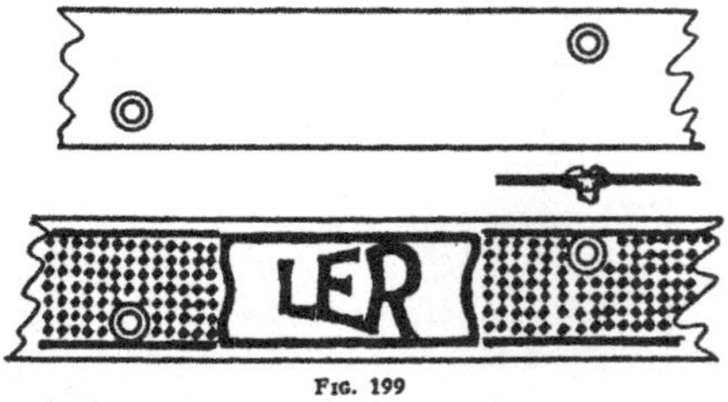

Fig. 199
Scuff plate of aluminum with etched design or car name

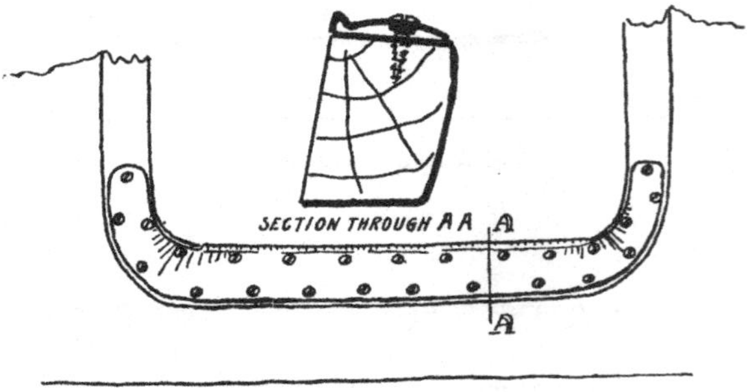

Fig. 200
One of the best designs for a scuff plate, with section through A A.

struction or else attach it to the wood raiser. In Fig. 193 is shown a section of a typical design; this can be made from No. 18 USS gauge.

Cushion Retainer. In the design under consideration the seat frames are

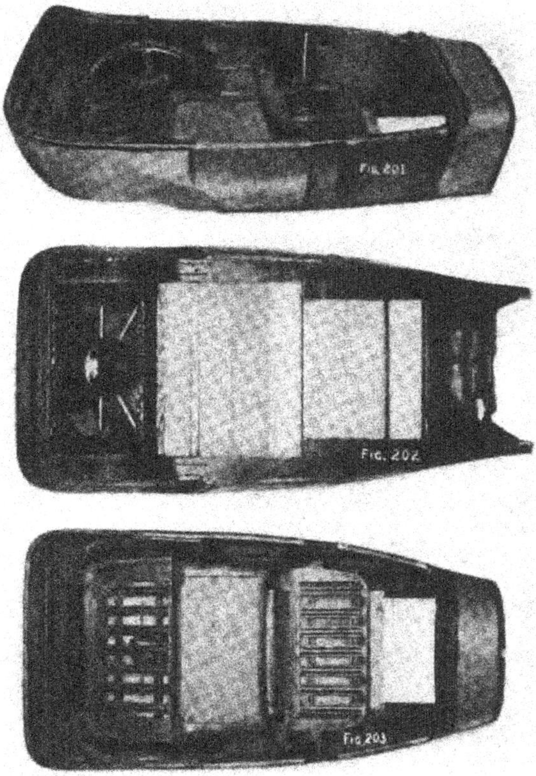

FIGS. 201, 202, 203
Example of all steel body construction (Edward G. Budd & Co., Philadelphia, Pa.)
Shows different views of same body so that the construction can be easily seen.

made of wood and it is necessary to attach a cushion retainer to the edge of the frame to keep the cushions in position. The simplest construction is to use a flat piece of round edge strip steel, fastening it to the frame by wood screws, as shown in Fig. 194. Another design, shown in Fig. 195 for a cushion retainer, is made from No. 20 USS gauge sheet steel. This makes a strong and nicely finished retainer, but is more expensive to manufacture. When the heel plate is made like that shown in Fig. 193, the cushion retainer is made by forming in the plate.

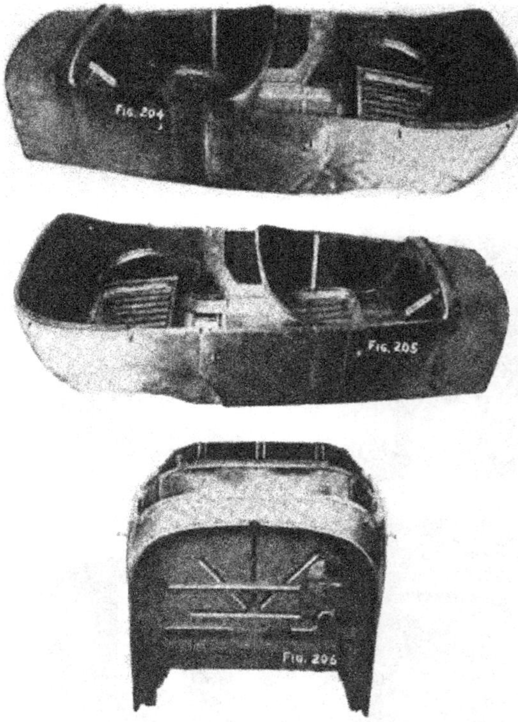

Figs. 204, 205, 206
Examples of all-steel body construction (Edward G. Budd & Co., Philadelphia, Pa.). Three views of the same body.

Scuff Plates. Sheet metal plates are attached to the tops of the threshold around the door openings, to cover the flanges of the panels where they turn over on to the wood. These are termed, "scuff plates" because they are also a protecting plate, as the feet catch on this part of the body when the passengers get in and out of the car. As there are a variety of ways of designing a scuff plate the several types in general use will be described. The cheapest design is to make them of sheet steel and nail them on, painting them the same color as the body. This is shown in Fig. 196 and section 197. It will be noted that the plate extends well up to the door post on each side. The reason for this is so as to cover the metal flanges of the panels, as there is a tendency for the metal to crack where the curved corners are drawn in the dies, and this scuff plate will extend up past this corner and effectively cover any cracks.

The next thing in order is to make a flat plate of aluminum and polish it, attaching it by French head wood screws to the threshold as shown in Fig. 198. This same style of plate can be engraved with a pattern on it, and the car's name also can be added like that shown in Fig. 199, and still be further improved. In Fig. 200 is shown one of the best, if not the best, design for a scuff plate. This is made of aluminum or aluminum alloy and can be either plain, embossed or engraved. The principal feature about the design is the small ridge raised upon the inside. See the sectional view. This has a double purpose, the first to provide a scuffing ridge and the second is that the ridge closes up the gap or space between the bottom of the door and the threshold, making a much neater finish than the other type.

After looking over the sheet metal parts on an automobile body the question that presents itself is that inasmuch as there are so many metal parts the wood might as well be left out. But there is a long stretch between the all metal body and the wood one, because it requires a very skillful construction to make a metal sill and to stiffen up the door frames and posts without adding excessive weight. As a matter of reference the photographs from No. 201 to 206 have been added so that the general construction of an all steel body can be appreciated. It will be noted that on one body wood door frames are used, and also that on both bodies all the toe and floor boards are made of wood. There are also a certain number of wood inserts made for attachment of the trimming; apart from these few items the bodies are all steel.

CHAPTER XII

AUTOMOBILE HINGES

The hinges that are used for the doors of the automobile body can be divided into two types, the outside and the inside. The outside hinge has a projecting portion that is visible on the outside of the body, while the inside hinge has its working portion concealed inside the door and leaves nothing on the outside to be seen. This latter type is generally called a concealed hinge and is used extensively on all modern bodies. These two types of hinges are made in a great variety of ways, and there are none that can be called the standard type.

A great deal of energy and money has been spent trying to evolve a hinge that will be very simple, efficient and cheap, and a number of patents have been granted for different designs that have been developed. A number of the different types that have been marketed and constructed will be described in this chapter, so that the work that has been done in this field can be reviewed and this may assist the designer should a new design of hinge be contemplated.

The first hinge used on the automobile was the type that had proven satisfactory on horse-drawn vehicles. This was of the flat butt style that is used so extensively on doors of all kinds in buildings and houses. This style hinge shown in Fig. 207 has proven very satisfactory and is used today on a great number of cars. From a mechanical standpoint they are very efficient, but from a viewpoint of neatness in appearance they are decidedly wanting, as they project from the side of the body, making an unsightly appearance and breaking up the smooth appearance of the outside of the car.

In Figs. 208, 209 and 210 are shown an approved design of flat butt. This is so shaped that the notches in the knuckles are concealed and make a better appearance than the type shown in Fig. 207. Apart from the above consideration of the outside hinge it has one advantage and that is that it permits the door to swing out from the body and keeps the top of the door level with the top of the body, while a concealed hinge allows the door to drop as it swings open. This condition is best illustrated by the sketch in Fig. 211 and Fig. 212. Fig. 211 shows the outside hinge swinging the door square, and Fig. 212 shows the door using a concealed type hinge and swinging out and dropping down. In order for a concealed hinge to allow the door to swing square the pillar would have to be made heavier so the hinge could swing as shown in Fig. 213. This condition is shown exaggerated to illustrate the description. It will be noted in these three Figs., 211, 212 and 213, that the pins in each type of hinge must line up, in other words,

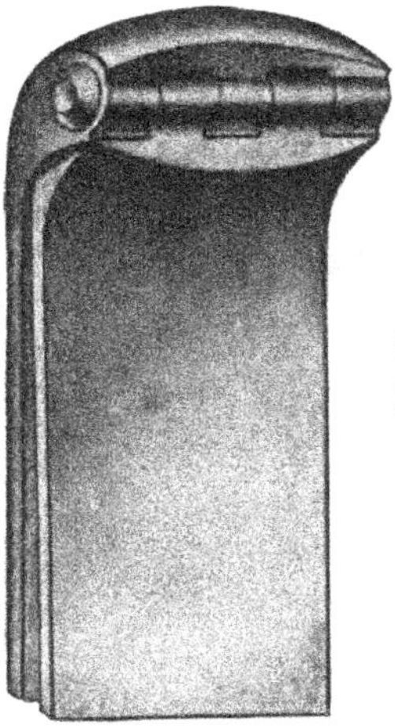

Fig. 207
Concealed Joint Curved Pattern

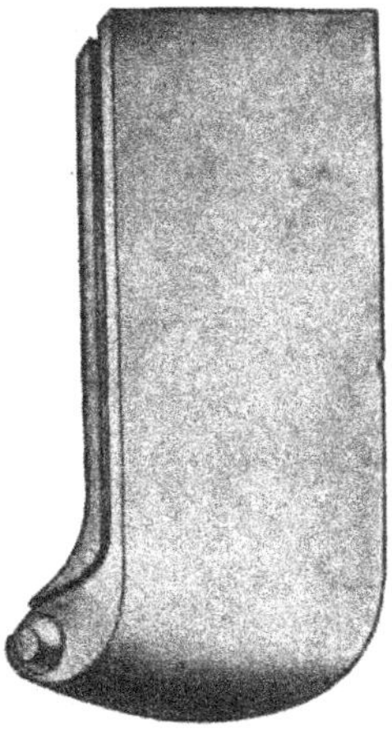

Fig. 209
Concealed Joint Curved Pattern

Fig. 208
Straight Pattern

Fig. 210
Concealed Joint Semi-Straight

Examples of curved and straight door hinges (Joseph N. Smith & Co.)

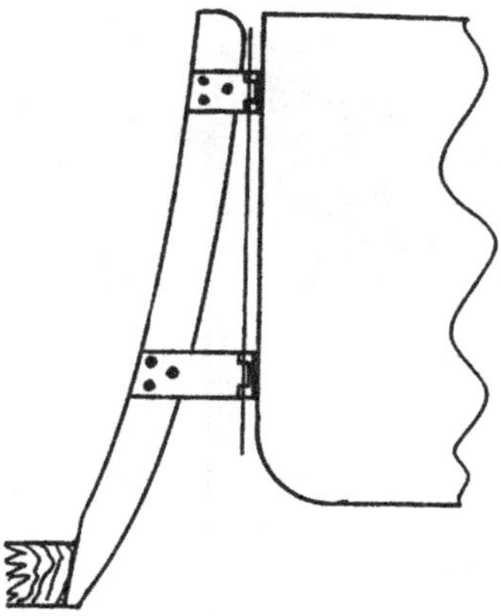

FIG. 211
Outside hinge swinging the door square

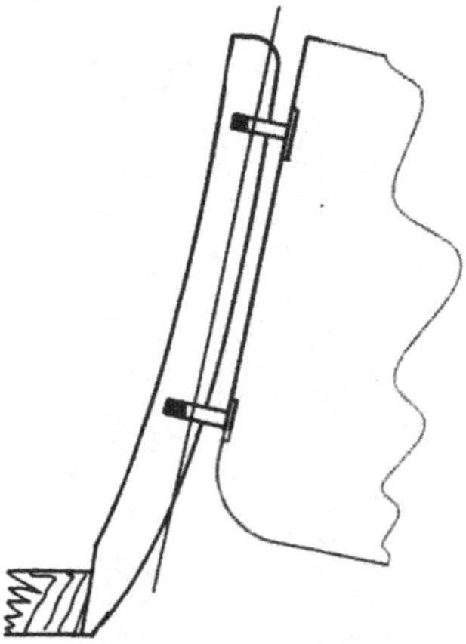

FIG. 212
Concealed type hinge swinging the door outward and slightly downward

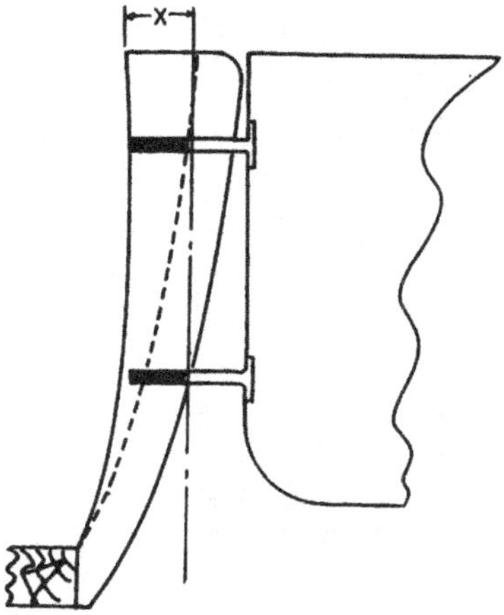

FIG. 213
Concealed hinge on heavier pillar, swinging the door square

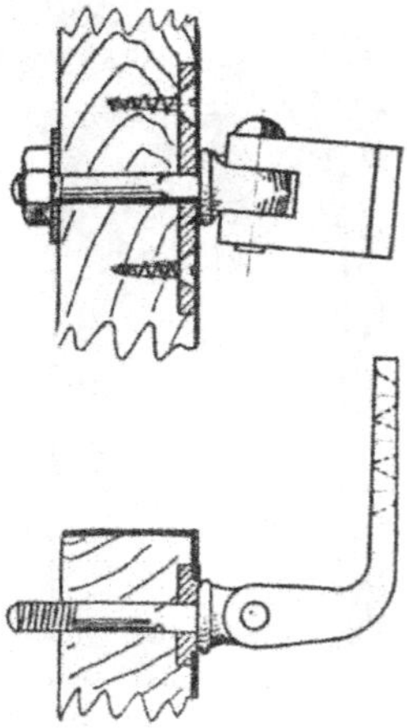

FIG. 214
Knuckle type hinge. With this style hinge the door is not readily removed

a center line must go through the points in which the door pivots; otherwise the door cannot swing.

The distance marked X in Fig. 213 will depend upon the amount of turn-under in the body, and in almost any case, even with a moderate turn-under, the post

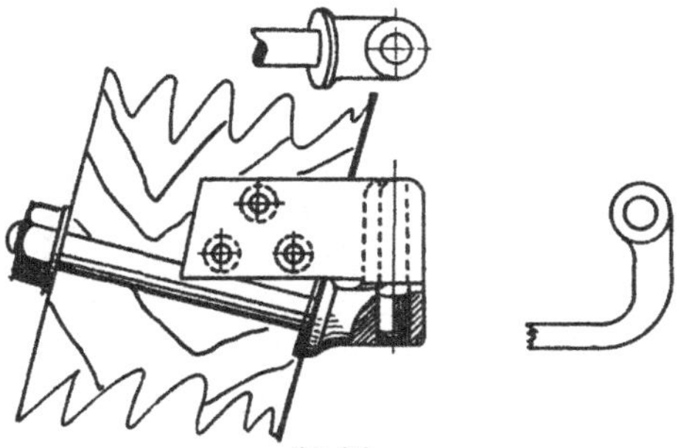

FIG. 215
Similar in construction to Fig. 214, but the two parts, instead of being riveted, contact on one face.

would have to be made too thick to be practical. A door swinging out square and level looks better than the dropping door, but the principal advantage of the straight swinging door is that in fitting side curtain supports or a winter top, the door extension will work much better.

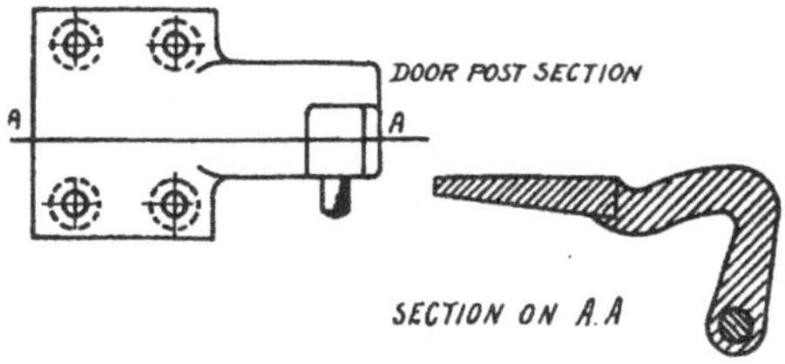

FIG. 216
Illustrating construction of a type of concealed hinge

In order to get the doors to swing square and yet not have an unsightly projection an outside hinge is often used at the lower section of the door. This is especially necessary on a closed type body where three hinges are generally used at each door. The two upper ones are concealed ones, while the lower is of the outside type.

To meet the above conditions the outside hinges shown in Fig. 214 and Fig. 215 have been developed and, as it will be noted, they are as neat as any projecting hinge can be. The hinge shown in Fig. 214 is of a knuckle style and does not have the advantage of being able to take the door off readily as the style shown in Fig. 215 does.

FIG. 217

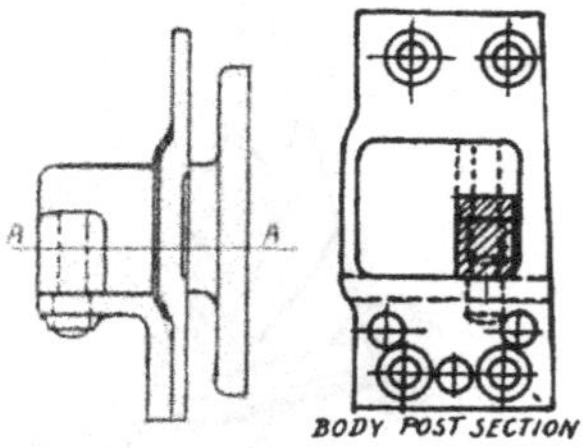

FIG. 218

Illustrating further the construction of a type of concealed hinge. See also Fig. 216 on page 105.

The hinge shown in Fig. 214 is made of two parts, a forged steel eye-bolt and a malleable iron bracket. This bracket is machined to fit on the bolt and is riveted in place so that it can turn freely. The hinge is held in place by a nut and washer on the end of the bolt, which is long enough to project through the door

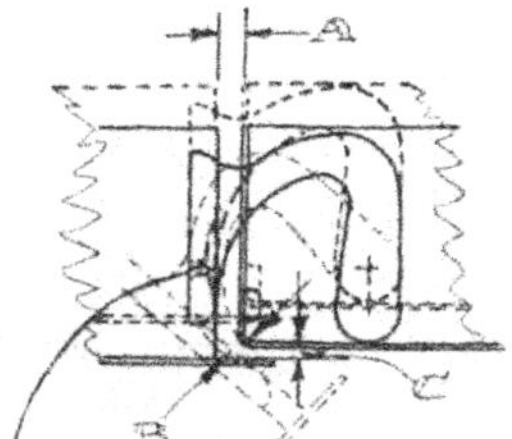

FIG. 219

Door post to illustrate the working clearance required

post. In order to prevent the bolt pulling down on the panel and throwing a buckle into it, a small casting with a circular machined boss is let into the post, the boss projecting through the clearance hole in the panel. When the hinge bolt nut is tightened, it pulls down on the iron bracket and so does not harm the panel.

The hinge shown in Fig. 215 is of similar construction to the first one, but the two parts, instead of being riveted together, contact on one face. The pin in the door section prevents them coming apart and allows the door to be taken off by lifting up on it so that the pin clears the hole in the bolt head. A plain steel washer is attached to the post under the metal, a hole being bored in the metal to clear the flange of the bolt. This eliminates the casting and answers the same purpose of preventing buckling.

The concealed hinge has been developed so that it can be very easily constructed and easily made. The principles upon which a hinge is laid out are as follows: It consists of two main sections called the door section and the body

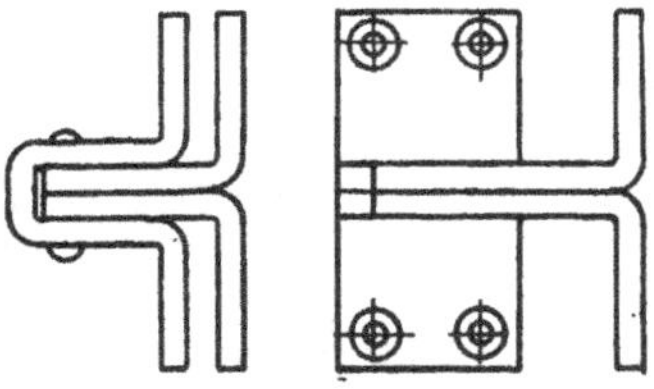

FIG. 220
The construction of the body section is of the simplest possible form, as shown here.

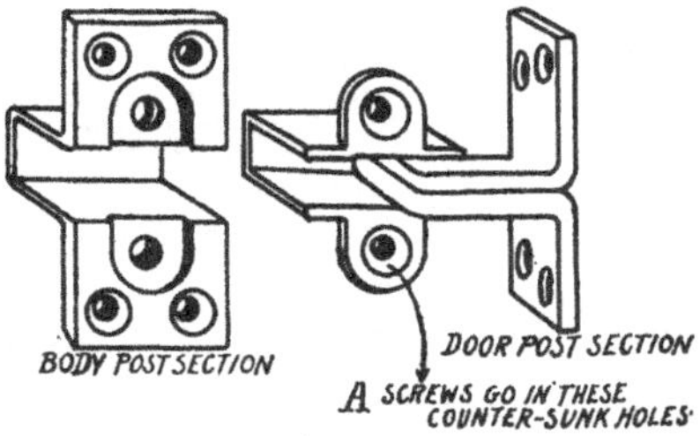

FIG. 221
A hinge that has the same door section as that shown in Fig. 220, but the body section is made so that when the two screws at A are taken out in the upper and lower sections, the door can be removed

section. The door section is an L-shaped section with a flange to attach to the door. This section swings or pivots on a pin that is either integral with the door section or the body section. See Fig. 216 and Fig. 217. The body section consists of a suitably-shaped bracket that can be secured to the post and provide the means to support the door section. See Fig. 218.

To illustrate the working clearance required, a door post is shown in Fig. 219. Two sections are shown, one for the upper hinge and the other for the lower one. The lower one is 2 inches back of the upper one, owing to the turn-under of the body. By measuring the distance between the inside surfaces of the two sections the angles that the door will make when it is open can be obtained. The principal considerations in laying out a hinge, are the door opening, panel and clearance between the flange of the door and the shroud panel. These points are marked A, B and C in Fig. 219. In this layout the door opening A is 5/32-inch, the corner clearance B, 3/32-inch, and the door flange clearance C, 3/32-inch. The corner clearance at the hinge is ample, but if a point is taken between the two hinges where the body sweep is at its maximum, there is only 1/16-inch amount of

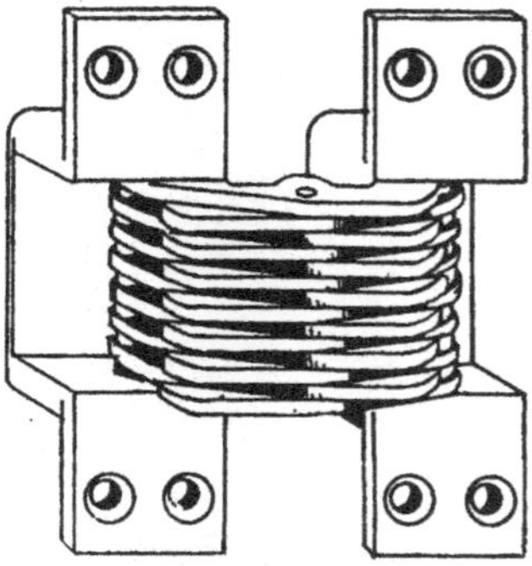

Fig. 223
Popular type of concealed hinge which is close working and very strong. (Soss Mfg. Co., Brooklyn, N. Y.)

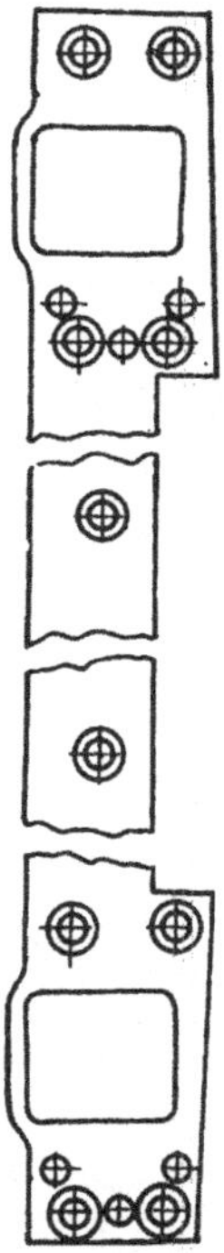

Fig. 222
Body section that has both the upper and lower section brackets attached to it

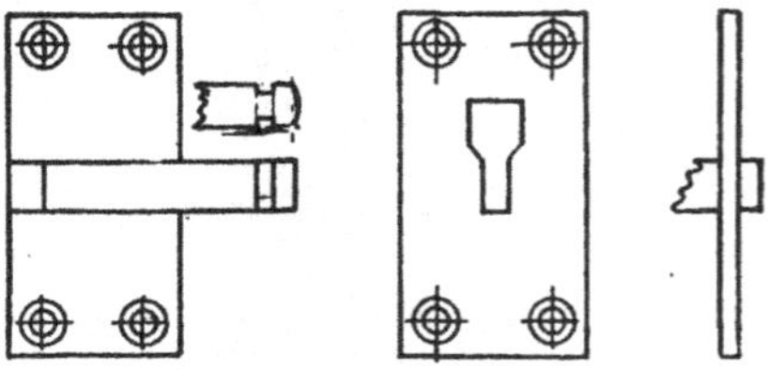

Fig. 224
A hinge which reverses the practice illustrated in the other designs here considered

clearance. In order to take care of this condition a part of the door frame is machined out as shown, so allowing the hinges to be made with a minimum amount of swing.

The construction of the body section is the simplest possible form as shown in Fig. 220. This is made from a piece of strip steel formed up as shown, and the door section which is riveted to it is also made of sheet steel as shown. In order to set this hinge either the door section is attached to the door and then the body section is attached, or vice versa. The body section is set in the body and then the door it attached. These operations are rather awkward and do not make a good

Fig. 224 A
Parsons patented concealed hinge. (Parsons Mfg. Co., Detroit)

assembling proposition. In case the door has to be removed it is necessary to take out the screws and bolts in either of the sections, which takes a lot of time. When the door is to be replaced it has to be readjusted, as the hinge cannot be set accurately. From the foregoing it is seen that some means should be provided to permit the door to be easily detached and set back again.

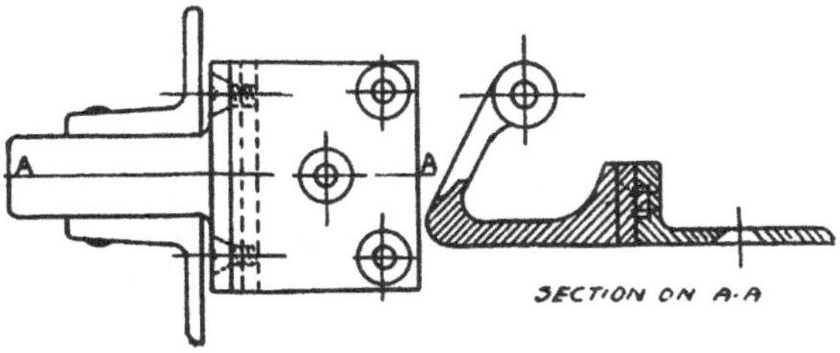

Fig. 225
Hinge made of malleable iron. The door is removed by taking out two screws in each door section

In Fig. 221 is shown a hinge that has the same door section as that shown in Fig. 220, but the body section is made so that when the two screws shown at A are taken out in the upper and lower sections the door can be removed. The recess in the door section insures that the door will go back in the same place without any trouble. Both of these hinges in Figs. 218 and 220 are substantial when constructed properly, and they are also economical to manufacture.

In Figs. 216 and 217 are shown designs of hinges that permit the door being taken off without having to remove any screws. The door section has a pin secured in it. This pin sets into a hole in the body section, as shown in the illustration. The dotted line shows the position the door section takes by lifting up

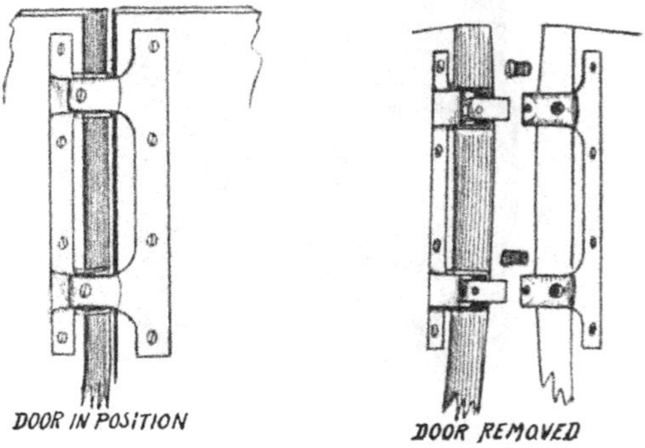

FIG. 226 FIG. 227
A sheet steel hinge with door and body sections in one assembly. Door shown in position in Fig. 226 and removed in Fig. 227

on the door, and which disengages the pin from the body section, permitting the door to be pulled right out. The door section is made of malleable iron and the pin on which it works is of steel, pressed into the iron. This door section sets on the side of the door frame, and it is bolted on, the bolts going right through the

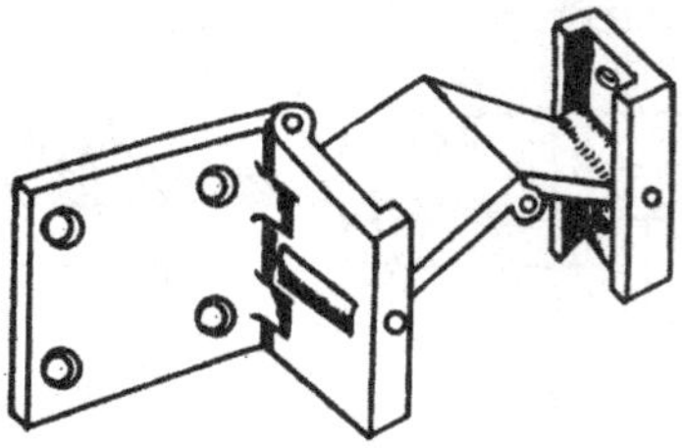

FIG. 228
Concealed type of hinge for doors that have large turn-unders

door frame. In Fig. 216 is a door section of similar design, only the flange is so located that it fastens to the back door frame. With these parts wood screws have to be used to fasten it to the door, as bolts cannot be used. In Fig. 222 is shown a body section that has both the upper and lower section brackets attached to it. It is made of steel plate and has malleable iron brackets riveted to it. The advantage of having the body sections all in one is that they are both bound to be accurately located when they are fitted to the body. In Fig. 223 is shown a hinge constructed from two malleable castings. This is a very simple construction but necessitates having the body section in two pieces. The design is substantially the same as the previous ones described. A Soss type of hinge is shown in Fig. 223. This is a very closely working hinge and very strong. Its construction is unique, as the working parts are built up of a number of pieces of steel so shaped as to give the desired movement. Fig. 224a is a Parsons concealed hinge, which is very largely used.

The hinge shown in Fig. 224 reverses the condition from the previous design described. On this design, the knuckle is part of the body section. The door section consists of a flat plate only. This plate has a keyhole-shaped opening in it which engages a groove cut in the edge of the body section knuckle. To remove the door it is lifted up until the knuckle pin disengages from the slot.

Fig. 225 shows a hinge made of malleable iron. The door is removed by taking out two screws in each door section. The hinges insure a correct alignment by the body section and door section, having a slot and a male portion projecting into the slot; this is shown in the section on A A. A sheet steel hinge is shown in Fig. 226. This has both the door and body sections tied together into one assembly, the object being to insure each section hinging with the other when they are fitted to the door and body respectively. The door is removed by taking out two small screws. The illustrations show the door removed in Fig. 227 and in position in Fig. 226.

A concealed type of hinge for doors that have large turn-unders is shown in Fig. 228. This hinge is especially adaptable to the doors of closed bodies and is designed to do away with the outside hinge generally used at the bottom of these doors. It will be seen that there are quite a number of parts to it and that it does not look as strong as an outside hinge. But with three hinges on the door this should stand up all right. Its design is very ingenious and as far as the author knows is the only hinge on the market to meet the condition that this one does.

AUTOMOBILE DOOR LOCKS

The door lock which is applied to the door of the automobile body has to meet a number of conditions in service that an ordinary type of lock which is used on the house doors, does not receive. Notwithstanding this, the principles of the automobile locks that are in general use are very similar to a common house lock. An

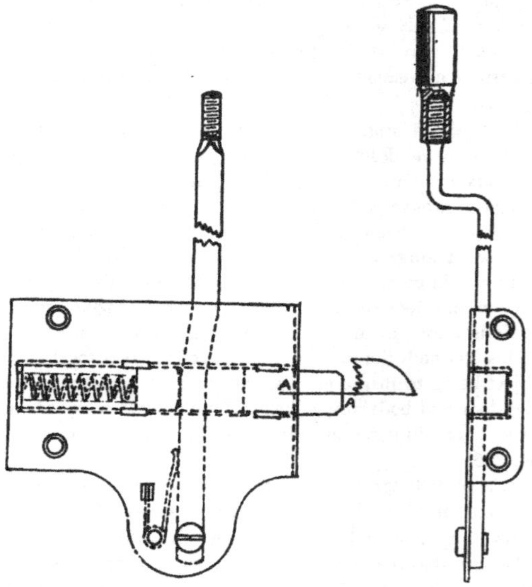

FIG. 229
One of a type of locks most generally in use

automobile door must be held securely by the lock, and not open under the strains and "twists" set up in the body when the car goes over rough roads.

The parts must be substantial and fitted so that they will not rattle. The lock must operate easily when opening and closing the door and at the same time should not have a projecting handle on the outside of the door or one on the inside that will catch in the clothing, or be liable to open the door by accidental contact. With the

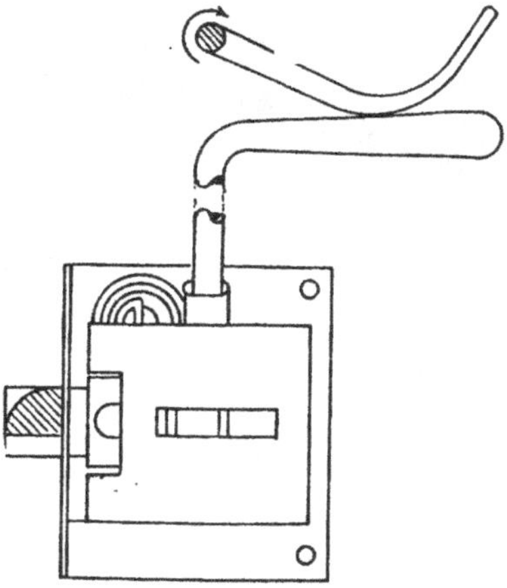

Fig. 230

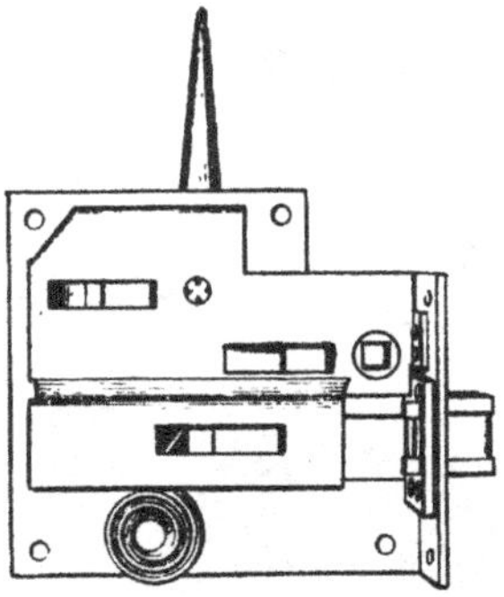

Fig. 232

Figs. 230 and 232 show typical rotary lever locks similar to the one in Fig. 229. The bolt, however, is operated by a rotating movement of the rod.

above conditions to be met, simplicity of design and ease of application on the car must be added, as the car builder must have a practical lock to conform with rapid production assembling conditions.

FIG. 233
Lock Lever Throw, (Jos. N. Smith & Co., Detroit)

FIG. 234
Lever lock for Coupe
(Jos. N. Smith & Co., Detroit)

FIG. 235

The Sears-Cross lock—a lock with an expanding latch
(National Seal Co., Inc., Brooklyn, N. Y.)

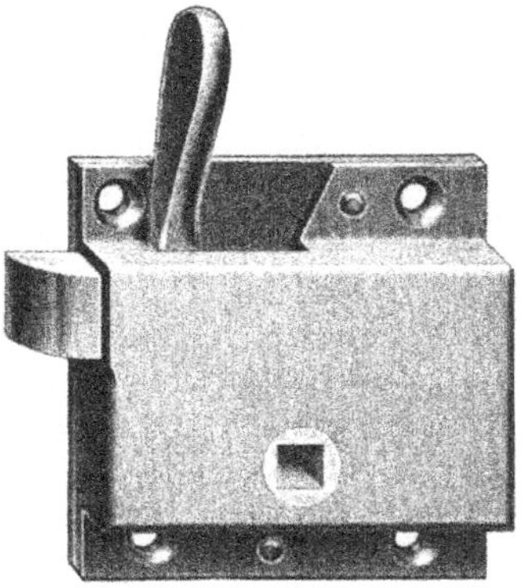

FIG. 236
A very compact and simple lock, well adapted to truck cabs.
(Jos. N. Smith & Co., Detroit)

With a view to improvement on the locks already in use, a large number of new designs have come on the market recently, but so far they have not gained very much favor. The type most in use is shown in Fig. 229. This style, with detail modification, is used by approximately a large percentage of the car builders. It

Fig. 237
Double action lock (Ferro Stamping & Mfg. Co., Detroit)

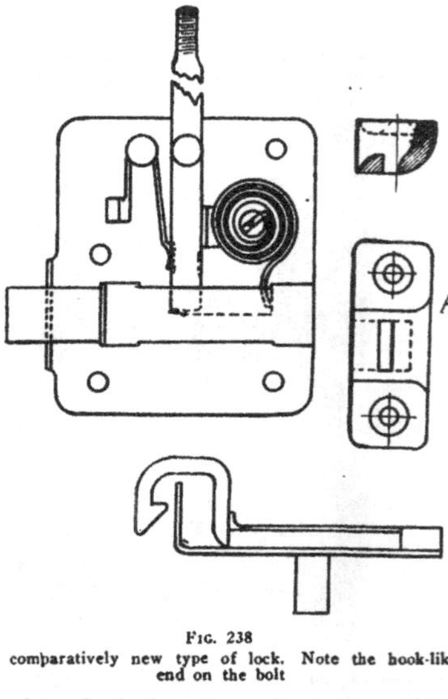

Fig. 238
A comparatively new type of lock. Note the hook-like end on the bolt

will be seen that it consists of a body or base plate with a sliding bolt operated by a pivoted lever. Two springs are used, one to throw the bolt out and the other to keep the lever stationary while the bolts work in upon striking the door post. It will be noted that if the lever were fast to the bolt it would move back when the bolt moved back, and if a person's hand were on the door near the lever while the door was being closed it would receive a sharp and painful blow when the lever flew back.

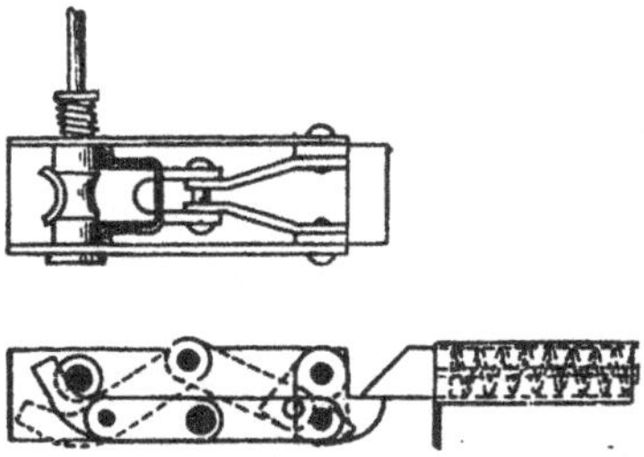

FIG. 239
A novel lock composed of a hinged bolt catch which is operated by two bars pivoted together

In Figs. 230 and 232 is shown a lock that has a sliding bolt similar to the design shown in Fig. 229. The bolt, however, is operated by a rotating movement of the rod. The curved handle shown is either pulled or pushed toward the center of the car. This pulls the bolt back and allows the door to open.

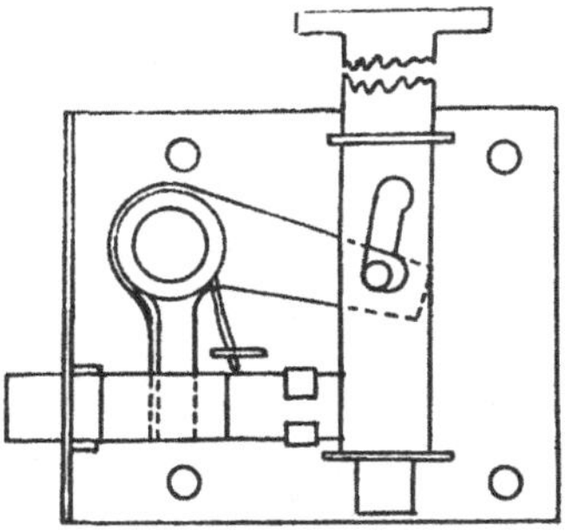

FIG. 240
A lock in which all the parts are of steel stampings with the exception of the bolt, which is of malleable iron

This method of operating a lock is preferred by some, and there are also movements worked out to operate the bolt by pulling up or pushing down, but these actions are not as popular as the standard sideways action of the door lock shown in Fig. 229.

When it is desired to open the door from the inside with a rotary movement and use the type of lock shown in Fig. 229 the lever can be cut off and a clip welded on so that a rotary handle like that shown in Fig. 233 will engage it. A new development of the 229 lock is shown in Fig. 234. This construction is very

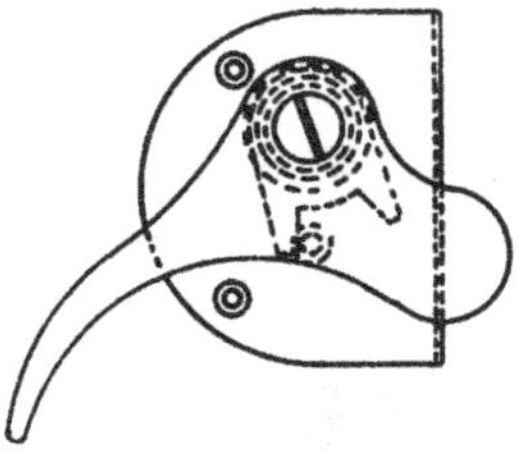

Fig. 241

Fig. 242

The lock shown in Figs. 241 and 242 is of a very simple construction, consisting of a base, a coiled spring, a stop plate and a handle.

simple. The lever engages a hook-end attached to the bolt which projects outside of the bolt housing. Provision is made for an outside handle by adding the die casting which engages with the lock bolt. A lock with an expanding bolt is shown in Fig. 235. This has a piece inserted in the end of the bolt so that when the bolt

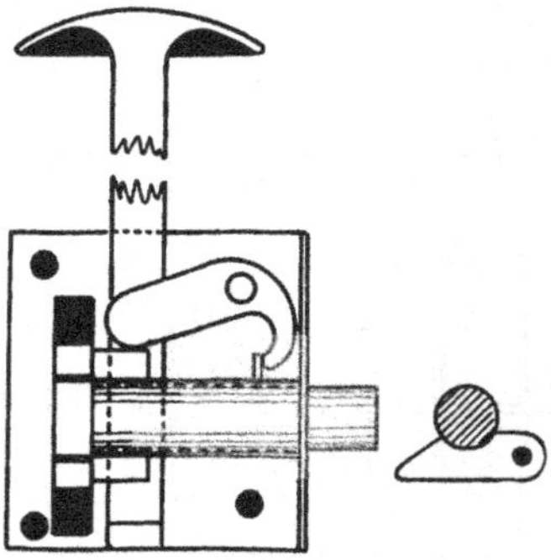

Fig. 243

A lock with a round bolt, operated by pulling up on the handle, which operates on the small lever, which in turn draws back the bolt

engages with the body strike the inserted piece projects out like a tongue and tends to pull the door tighter and prevent rattles. A very compact and simple lock for truck cabs is shown in Fig. 236. One of the simplest of automobile locks is the adaptation of the lock 229 so it can accomodate outside door handles. By examining Fig. 237 it will be seen that the lock construction is just the same as 229, only a tumbler and housing are added. This addition gives all the conditions previously outlined and also allows the use of an outside door handle which will operate the lock with either a right or left turn. There are a number of door locks made which can be operated one way only, either a right or left turn.

There is a new design of lock that has several new points over the old types. It works on an entirely new principle; the bolt comes out upon hitting the strike,

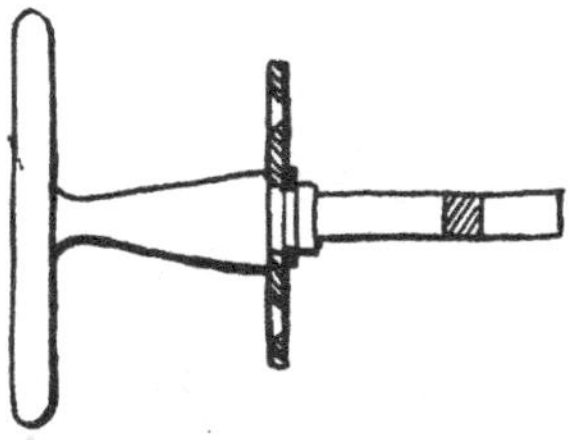

FIG. 244
Construction of a typical door handle

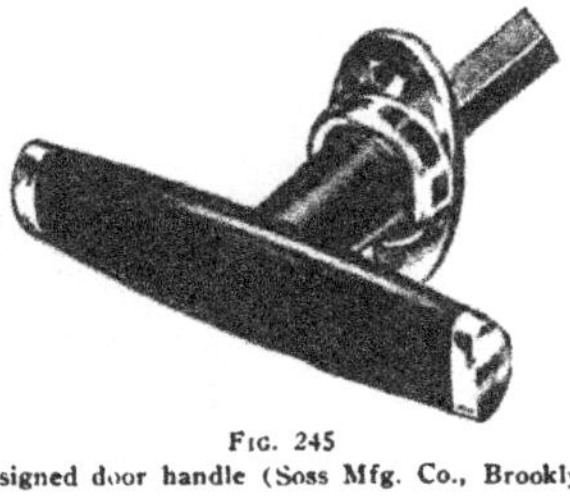

FIG. 245
A well designed door handle (Soss Mfg. Co., Brooklyn, N. Y.)

instead of going in, as it does in the type shown in Fig. 229. This lock is shown in Fig. 238. It will be noted that the bolt has a hook-like end. This hits on the strike shown at A, and when it passes the point on the strike it flies backward and catches securely in position. When the door works back and forth in service the bolt works with it. This gives a condition of engagement that will prevail no

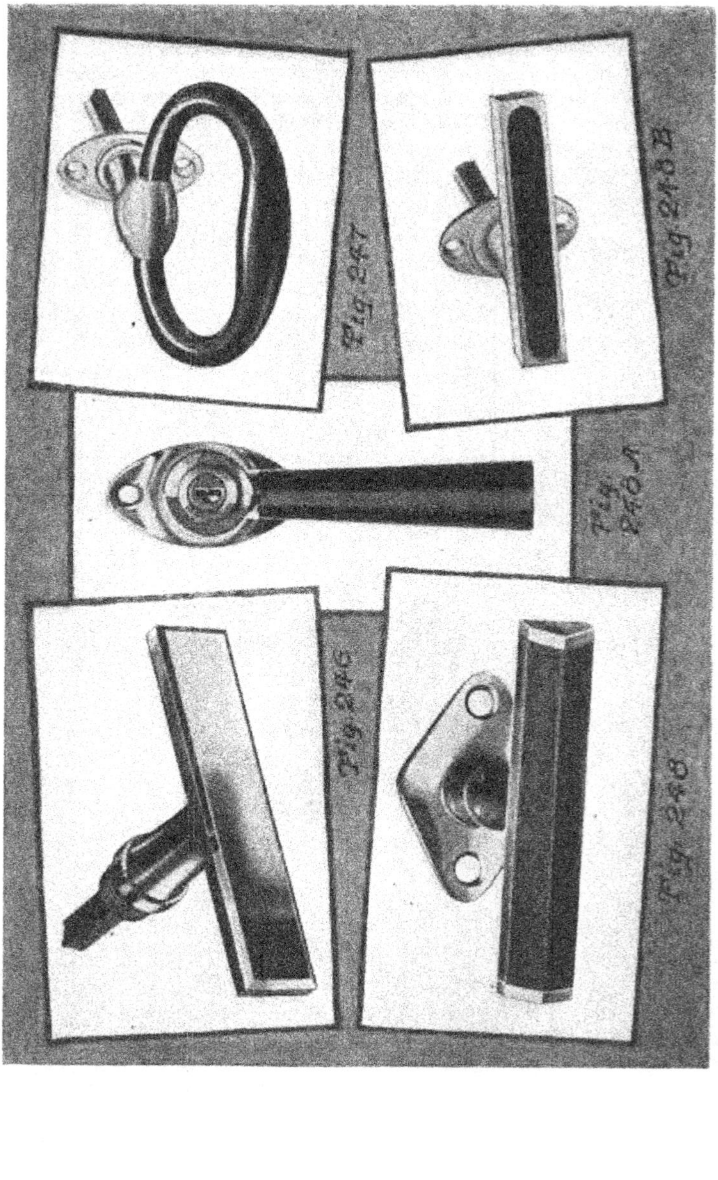

FIGS. 246, 247, 248, 248 A and 248 B illustrate typical door handles made by leading manufacturers. Fig. 246, English & Mersick; Fig. 247, Mitchell; Fig. 248, Ferro Stamping & Mfg. Co.; Fig. 248 A, Ferro Stamping & Mfg. Co.; Fig. 248 B, Mitchell.

matter how much the body twists and weaves; the door cannot open unless the lock is operated. The action of operation is a straight pull or push on the lever to open the door.

A novel lock design is shown in Fig. 239. This is composed of a hinged bolt catch which is operated by two bars pivoted together. When the bars are in normal position they lock the catch rigidly, and in order to close the door a spring strike has to be used. This is pushed in when the catch hits it and flies back when the bolt has passed the point, giving a state of engagement similar to the regular lock.

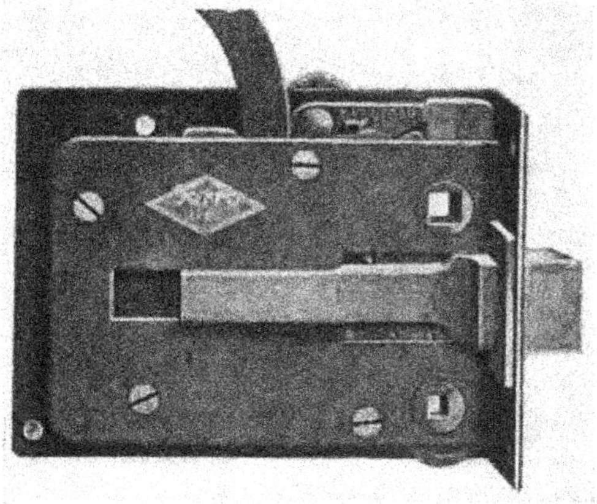

FIG. 249
This lock can be locked from the inside by turning the knob, or locked from the outside by the cylinder lock, which operates the same pawl as the knob does
(Ferro-Stamping & Mfg. Co., Detroit)

The position the catch takes when the door is closed is shown in the sketch. To open the door pressure is exerted on the pin A. Through a suitable button and connection it throws the bars out of line and permits the catch to swing back, as shown by the dotted lines; the door is then free to open. The spring shown on the top of the left-hand bar returns the catch to a locked position ready to catch when the door is closed.

In Fig. 240 is shown a lock in which all the parts are of steel stamping with the exception of the bolt, which is malleable iron. The bolt is of the regular sliding bolt type and it is operated by an upward pull on the handle. It will be seen that the bell crank is operated by the pins engaging on the bottom of the slot in the upper arm. The lower arm engages in the s'ot in the bolt and pulls the bolt back when the door is opened.

The lock shown in Fig. 241 and 242 is of very simple construction, consisting of a base, a coiled spring, a stop plate, and a handle. The handle forms the bolt and when the rounded end hits on the lock strike the handle is pushed back, allowing the door to close. The position the handle takes when the door is closed and locked is shown in the lower sketch. A lock that has a round bolt is shown in Fig. 243. This bolt is operated by pulling up on the handle which operates the small lever, and this in turn draws back the bolt. The style of strike used with this lock is shown in the small section at the right of the lock.

FIG. 250.

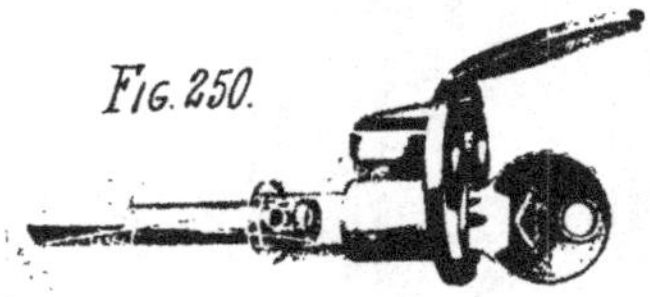

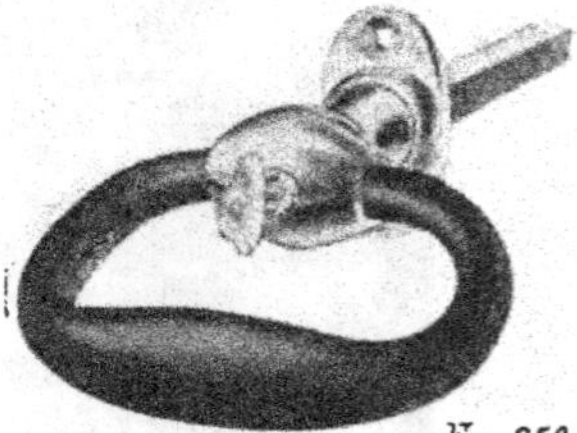

FIG. 250A.

Cylinder Key Lock (Ferro Stamping & Mfg. Co., Detroit)

Mitchell Locking Handle (Mitchell Specialty Co., Philadelphia)

Ferro Stamping & Mfg. Co.'s Key Lock No. 824, and Mitchell Locking Handle

The style or vogue of the automobile affects the style of door locks. At one time all cars had outside door handles and then came a demand for flush exteriors, so that all outside handles were eliminated. With the improvement of side curtains difficulty was encountered in getting in and out of the car unprovided with outside door handles so that the present vogue is to have an outside handle. Door locks which will accomodate these handles have to be used. The door handle consists of a bar with a square ended shaft in it, which must be long enough to reach the tumbler socket. An extension plate is fastened to the handle by a lock plate or washer so the handle can not pull off. The construction of a door handle is clearly shown in Fig. 244. The designs of door handles vary considerably and there are a great number of different designs now carried by the automobile hardware manufacturers, some of which are shown in Figs. 245, 246, 247 and 248.

The closed car door lock has the same basic principle of construction as the open body lock, Fig. 229, has, only it is much stronger and inside and outside locking devices have to be incorporated. The lock shown in Fig. 249 can be locked

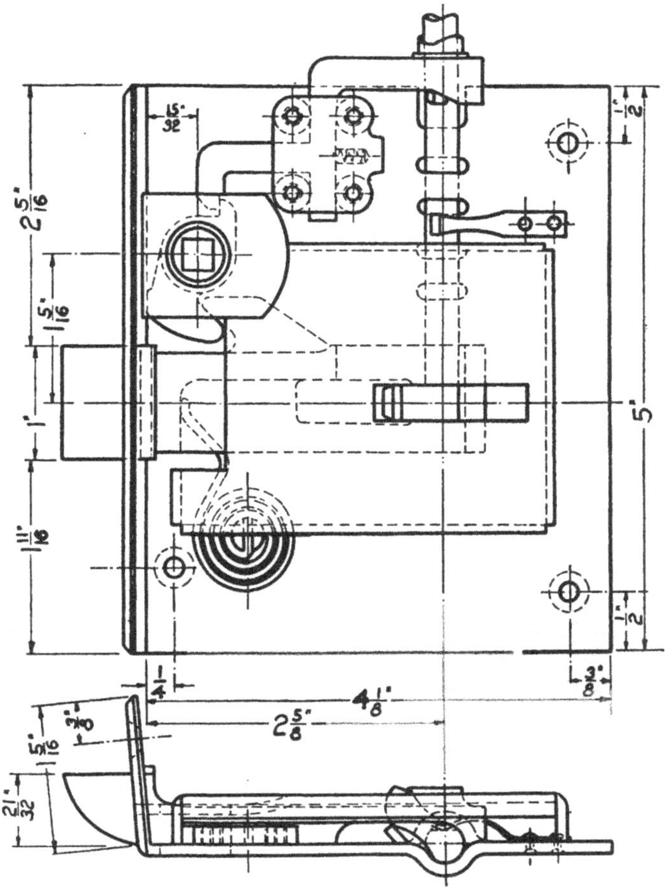

Fig. 250 B
Rotary lever lock handle with automatic release on the inside
(English & Mersick Co., New Haven, Conn.)

from the inside by turning the knurled knob or locked from the outside by the cylinder lock which operates the same locking pawl as the knob does. The illustration shows how the cylinder lock is attached. A cylinder lock is shown in Fig. 250. It will be noted that the shaft has a square end to engage the lock and that it is connected to the cylinder lock loosely, so it can take care of misalignment.

In 250B is shown a rotary lever lock with an automatic release on the inside, This unusual lock presents many advantages, and was first introduced to the trade in the winter of 1921-1922.

A number of different types of locks have been shown which cover a good many ideas that have been worked out in the endeavor to improve the lock design. Some of these locks are giving good satisfaction, some have not been so popular. The idea of giving this description of various types is to supply a record of the developments in this important fitting of automobile bodies.

CHAPTER XIV

GENERAL AUTOMOBILE HARDWARE

Under the heading of general automobile hardware will be grouped the door lock strikes, and as this chapter follows the chapter on door locks, the strike will be the first piece described.

LOCK STRIKES

The lock strike or "strike," as it is commonly termed, is a fitting that is provided to make the operation of closing the door as smooth as possible. When the

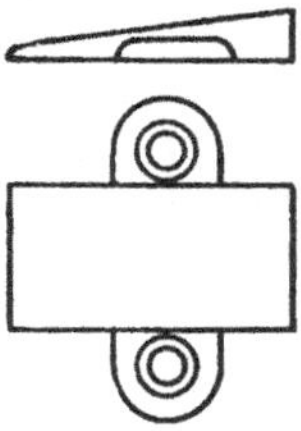

Fig. 251
Where possible a long strike is used

Fig. 252
Sometimes a shorter strike is necessary

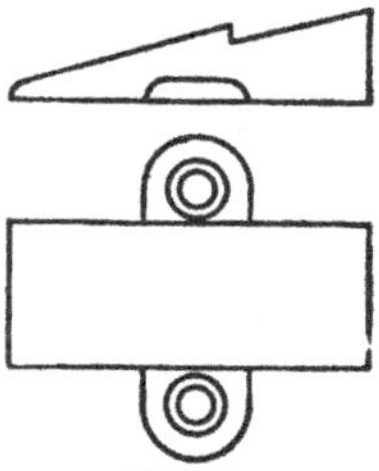

Fig. 253
Strike which provides a safety catch

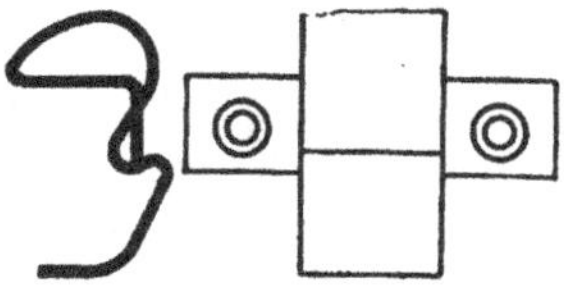

Fig. 254
Typical spring catch

door closes, the lock bolt has to be thrust back, and in order to do this easily a suitable bevel is provided that will make a sliding contact. The length of the sliding face of the strike is governed by the thickness of the door frame and lock pillar. When it is possible, a long strike like that shown in Fig. 251 is used, but

Fig. 255

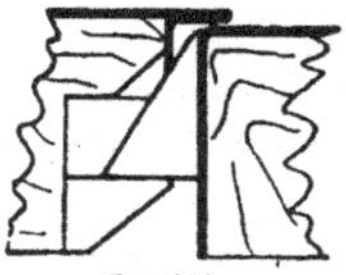

Fig. 256

Fig. 257

Figs. 255, 256 and 257 illustrate the three well-known methods of fitting the door into the frame

Fig. 258
An early bumper design

Fig. 259
First door condition

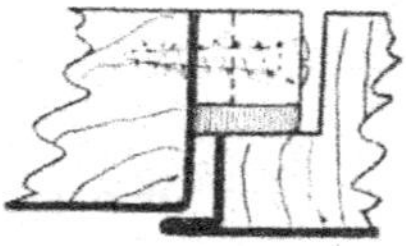

Fig. 260
Second door condition

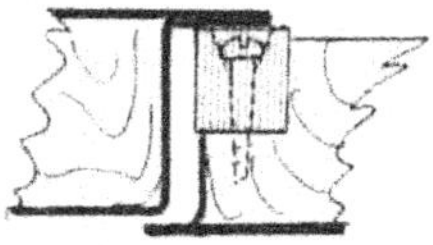

Fig. 261
Third door condition

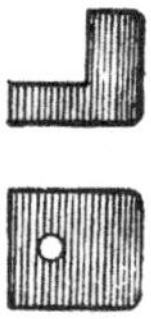

Fig. 262
A standard bumper

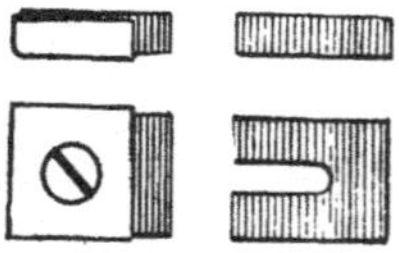

Fig. 263
Fig. 262 in position

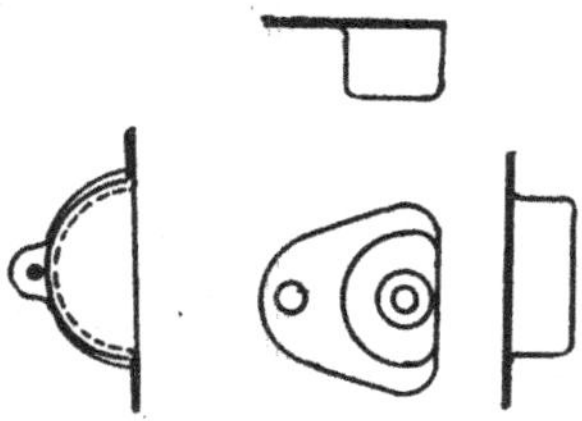

Fig. 264
Bumper used without door rabbet

Fig. 265
Ternstedt bumper

FIG. 266
A spring type of bumper

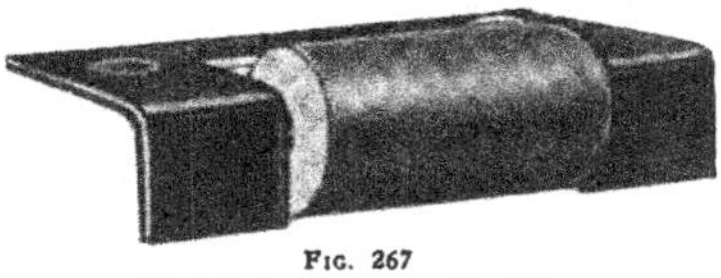

FIG. 267
Mitchell door bumper and retainer

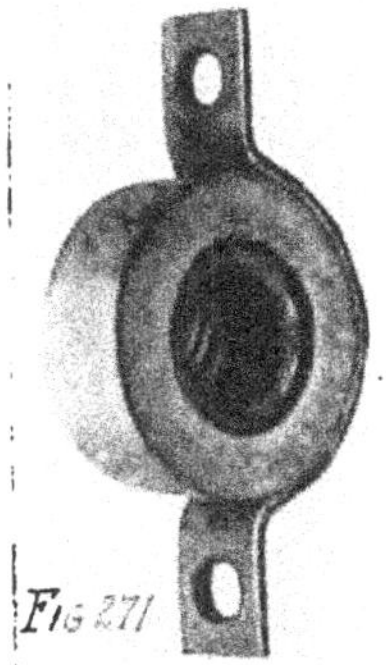

FIG 271

Mitchell Combination dovetail door bumper

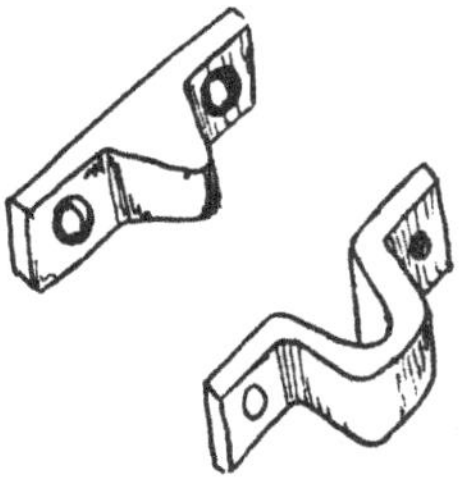

FIG. 268

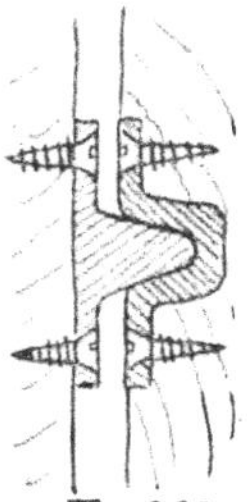

FIG. 269

Typical design of dovetail, and cross-section of same, taken through the door post, dovetail and door frame.

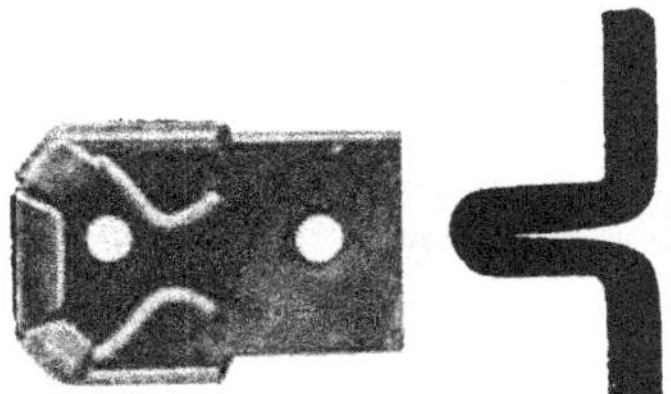

FIG. 270

Ferro Stamping Co's. dovetail door bumper

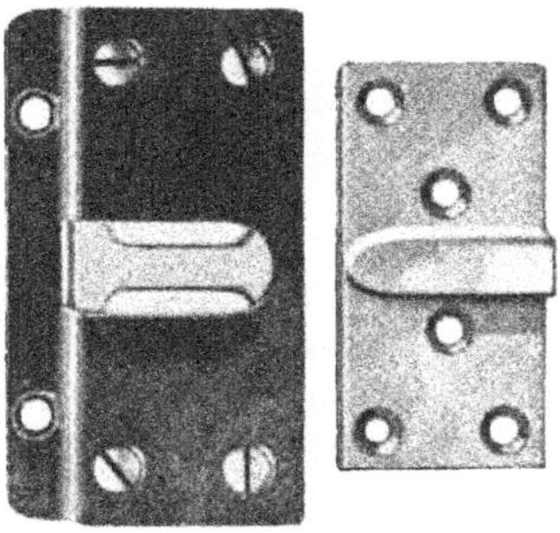

FIG. 272
Ternstedt "Ideal" dovetail door bumper

where the limitations of the design make a small strike necessary, a type similar to that shown in Fig. 252 is employed. In the first case the strike is used with a lock that has the end of the bolt rounded off, and in the second case the lock bolt has a straight bevel, the strike having its front edge rounded off.

FIG. 273.
A successful design of dovetail

In Fig. 253 is shown a strike that provides a safety catch. This holds the door in a partially closed position on the first notch in case the door is not swung to with sufficient force to insure it closing tight. Several different types of spring

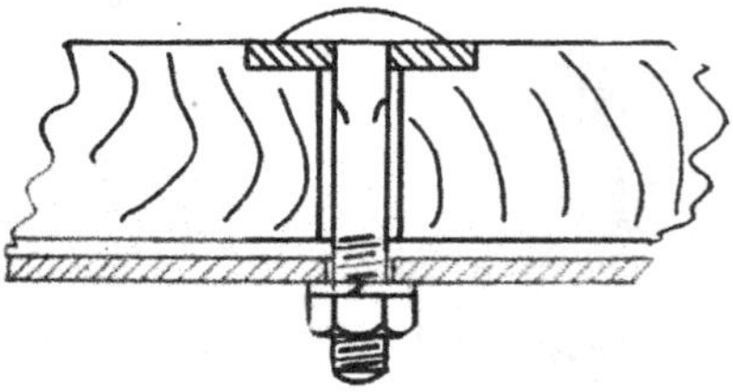

FIG. 274
Body bolt washer in position

strikes have been worked out, one of which is shown in Fig. 254. This is made from a strip of steel formed up and hardened and tempered. The drawback to this type of strike is that it is very liable to break under the hard usage that it gets,

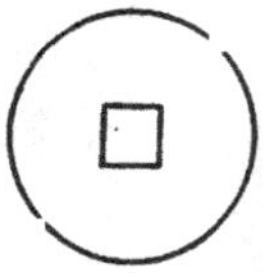

FIG. 274 A
Illustrating the conditions described in the text on page 133

and if the hardening and tempering is not very carefully done it will prove unreliable.

When considering the type of strike to be used the construction of the door and door frame will govern the design to some extent. Apart from the varying thickness of the posts there are three methods of fitting the door into the frame. The first (shown in Fig. 255) is to let the door set flush in the opening with just sufficient clearance around it to allow it to work. The second condition is to have a flange on the door panel which will conceal the opening of the door. The body post will have a rabbet in it, as shown in the section sketch in Fig. 256. The third type eliminates the rabbet, as shown in Fig. 257. With this latter type is generally used the strike shown in Fig. 251, while in the second type the strike shown at Fig. 252 is used. The first type is not favored very much, and takes a modification of the strike used on the third style.

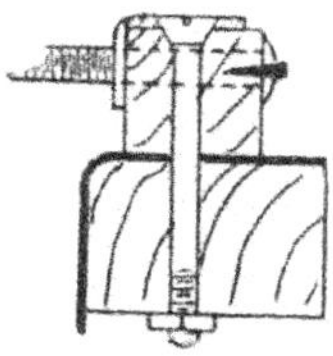

Fig. 275

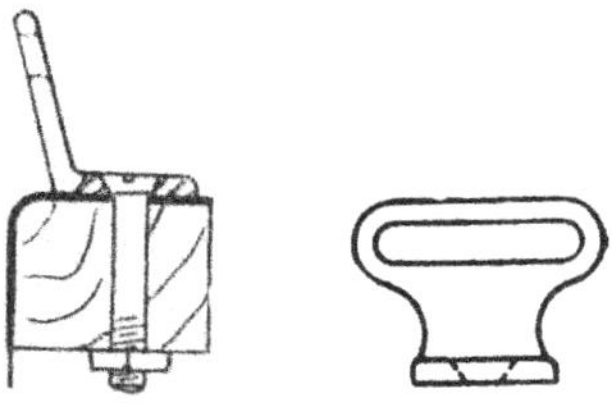

Fig. 276

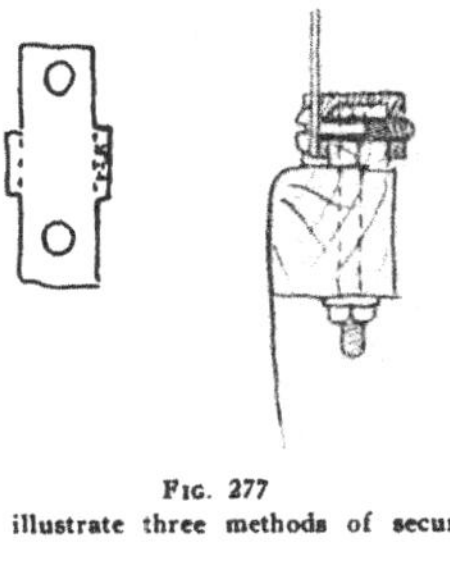

Fig. 277

Figs. 275, 276, 277 illustrate three methods of securing the top straps

DOOR BUMPERS

The door bumper comes next in order and consists of a means for regulating the distance the door shall swing in when the door is closed. When the door is slammed shut a fairly rigid contact has to be provided to insure the door closing to a definite position each time. This is especially necessary when a flanged door is provided, as the flange would strike on the body every time the door was closed if a stop were not provided. The early method of doing this was to set a metal plate on the body posts at convenient points and at the points of contact on the door were set screws which could be turned in and out as desired, to provide the right degree of engagement. This arrangement gave a hard closing door, as the shock of closing the door was taken by solid surfaces. In order to eliminate this condition rubber stops or bumpers were used in place of the solid contact.

The earlier design of rubber bumpers consisted of a small circular pad of rubber with a nail in the center as shown in Fig. 258. From this there has been

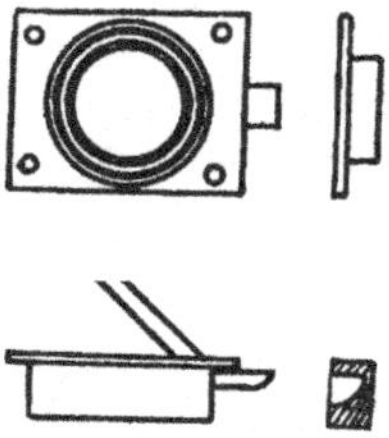

Fig. 278

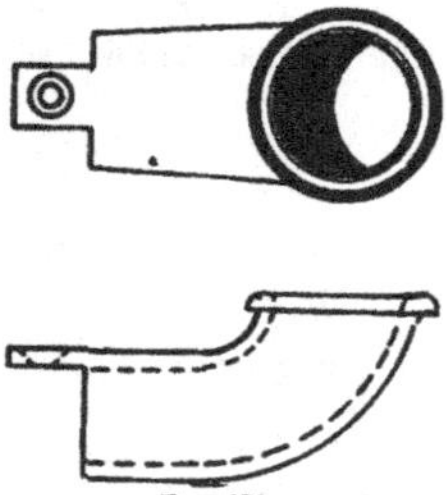

Fig. 279

a great variety of different forms of bumpers devised. A number of designs have been developed that have springs in them to take up the shocks. There are some that have both springs and rubber, but the most satisfactory bumpers that have been devised are of rubber. The bumper design is affected by the door post construction in a similar manner to the strike. The first door condition would

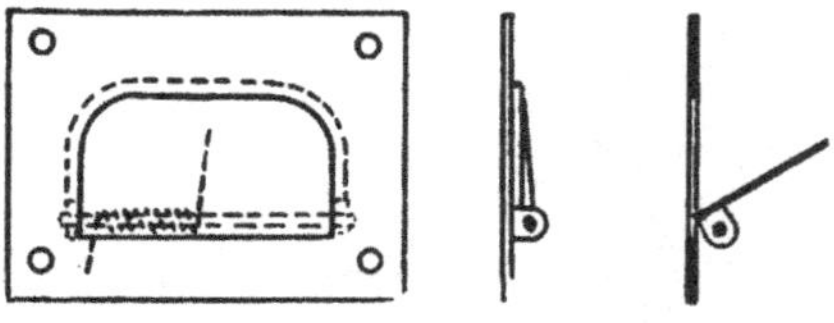

Fig. 280

The illustrations on this page show three different kinds of finger plates and hole brackets that are in common use

need a bumper similar to that shown in Fig. 259, while the second would take one like that shown in Fig. 260; in Fig. 261 is shown one that could be used in the third case. In the following will be shown a number of bumpers that will be of interest to the designer.

In Fig. 262 is shown a standard design that is used very extensively. This sets into a recess provided in the door rabbet and is held in place by a wood screw.

It has a good area of contact and makes a very satisfactory bumper. This is shown in position in Fig. 263. The design shown in Fig. 263 meets the same condition as the previous one, only the rubber is made adjustable by the slot in it, the screw holding it fast in the desired position. The bumper shown in Fig. 264 is used on a door without a rabbet and consists of a semi-circular piece of rubber. The sketch in Fig. 259 shows how it sets when the door is closed. In Fig. 265

FIG. 281
A conventional design of foot rest with nickel plated rod and nickeled or enameled end brackets

is shown a bumper that reverses the conditions just cited; instead of the rubber bumper going into a steel pocket the metal male section goes into a semi-circular rubber. This makes a very effective non-rattling bumper. A spring type of bumper is shown in Fig. 266 and in Fig. 267 is shown a round rubber bumper set in a permanent bracket, the idea being that the rubber can be turned to present a new face should any section of it become worn.

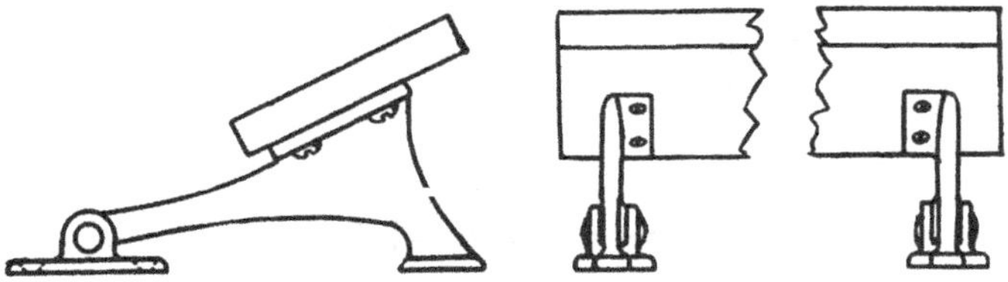

FIG. 282
Foot rest made of two end brackets and a center board, covered with a suitable material

THE DOVETAIL

The dovetail is provided on a door to insure its closing in the right position, and to relieve the strains that would be imposed on the hinges by the weight of the door distortion of the body or by passengers leaning or sitting on the door. When the hinges have been strained the door will drop on opening and then be hard to close. The dovetail consists of two sections, a male and female. They are generally made of composition metal, but can be made of malleable iron or steel.

In Fig. 268 is shown a typical design of dovetail that is used very extensively. The female section is attached to the door and the male section to the door post. The cross section shown in Fig. 269 is taken through the door post, dovetail and door frame. The male section is V-shaped and is so set on the door that it will set tight into the female section, which has a V-shaped groove to correspond. A

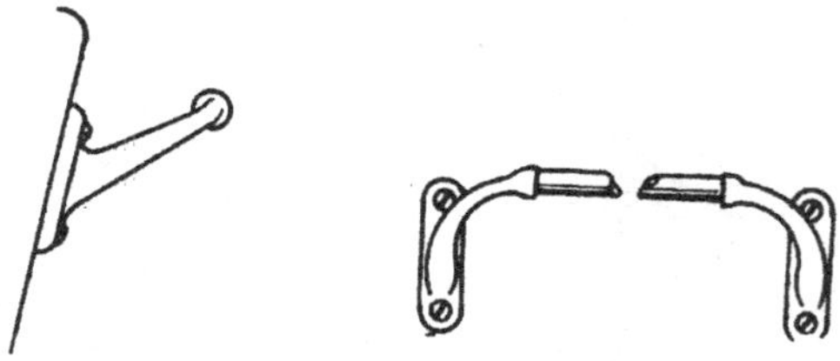

FIG. 283
A solid type robe rail that has proved acceptable

spring type is shown in Fig. 270. This permits very easy engagement and compensates for distortion in the body. The female section of this consists of a sheet steel retainer which holds in position a formed steel spring. This spring is hardened and tempered. The male section is formed up out of sheet steel and into a V shape. A bumper can be designed to act as a dovetail, like that shown in Fig. 271,

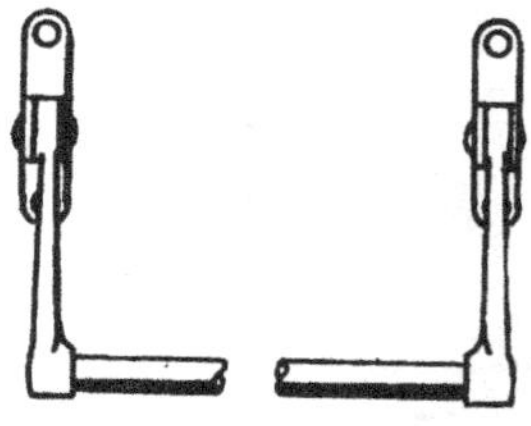

FIG. 284
This robe rail folds against the back of seat as shown in Fig. 285

the bumper may be formed like that shown in Fig. 265, which makes the bumper perform the function of a dovetail as well as of a bumper. A design of this type is very effective when it is new, as it prevents any rattling of the door, which is liable to occur in the type shown in Fig. 269. To make this bumper effective a liberal-sized rubber must be used.

On closed cars a heavy dovetail is used. A very successful design is the dovetail shown in Fig. 272. This consists of a die-cast housing with metal-protected rubber insets. A metal housing goes over this and makes a finish as well as holding the parts in the housing. The male section is cast metal and shaped as shown.

BODY BOLT WASHERS

Body-holding bolt washers are a detail of interest, as it is very important to have the body fastened securely to the chassis frame. Carriage bolts, 3/8 or 5/16-inch diameter are mostly used; the 3/8-inch size is the better. It is necessary to have a liberal-sized washer under the head of the bolt so that it will not pull into the wood. A cheaply-manufactured washer can be made from 1 x 5/32-inch strip steel, about 1½ inches long, as shown in Fig. 273. This is used with a bolt that

FIG. 285

FIG. 286

Fig. 285—Robe rail folded against back of seat. See Fig. 284.

Fig. 286—Flexible robe rail which can be detached at one end and the robe placed upon it before it is buckled again.

fits the body hole properly, but if the hole is enlarged it is not very satisfactory. A very simple and satisfactory system is to have a 5/8-inch hole in the body and use a circular washer, Fig. 274A, with a square hole that will fit the head of a carriage bolt. The large hole in the body permits the lining up of the hole in the chassis with the body hole, while the washer prevents the possibility of the body shifting, as it fits the bolt. The condition outlined is shown clearly in section 274.

TOP STRAP FASTENER

Another detail that is very important, which at the same time appears to be quite trivial, is the means provided to attach the top strap to. The top imposes a great strain on the top straps in service, and if it is not secured to a substantial construction, something will tear loose. Three methods of securing the top straps are shown in Figs. 275, 276 and 277. The first, shown in Fig. 275 is used where no adjustment is provided in the straps, and consists of a steel bracket with an ear that is bent over to provide a support for the bolt. This bracket is attached to the body by two stove bolts that go right through the rails or rail, and has a plate and nuts underneath. This distributes the strain of the straps onto the body rails. The staple fastened over the head of the bolt is to prevent the bolt being accidentally pushed back into the trimming. If this happens, the trimming would have to be pulled off to get the bolt back again.

The second method, shown in Fig. 276, is adaptable to straps that have a buckle adjustment, as they can be threaded into the eye that is provided. This eye or loop is made from steel or malleable iron, and is held in place by a bolt or bolts going right through the body rails.

The third system, Fig. 277, is to use a plate similar to Fig. 275, only one ear of the plate is tapped out so that the screw can be turned into it. This does away with the necessity of using the carriage bolt and makes an easy assembly.

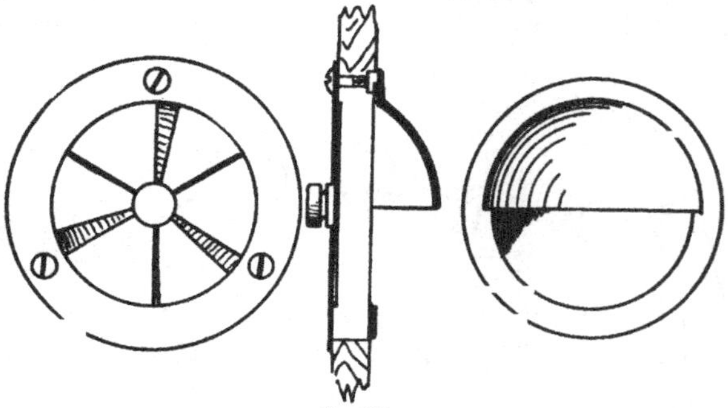

Fig. 287
Circular type of ventilator designed to attach to the shroud or cowl

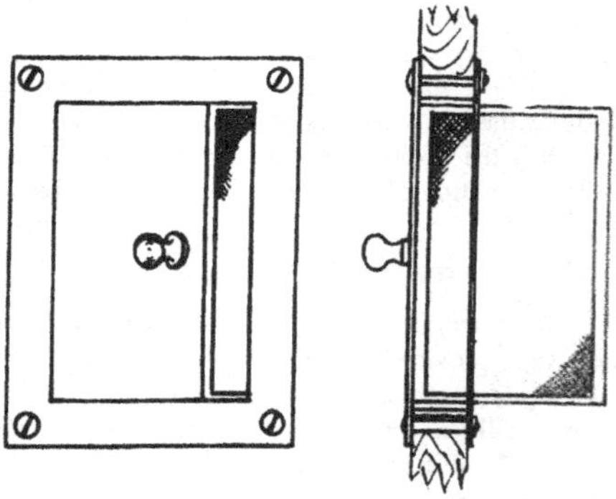

Fig. 288
Sliding plate type of ventilator designed for top of shroud or cowl

FINGER PLATES

On the majority of cars it is desired to have ready access to the mechanical parts under the floor boards in the front part of the car. In order to do this the one or two boards, as the case may be, are left loose in the body so that they can be readily pulled up. In order to raise the board a catch or hole is provided in the board, and in Figs. 278, 279 and 280 are shown three different kinds of finger plates and hole brackets that are in common use. In Fig. 278 a spring catch is shown that hinges to the position shown in lower sketch, and provides a means for pulling up the board. The bracket shown in Fig. 279 is made from a non-corroding alloy and is so shaped as to permit a finger being inserted into it. The curve is put in so that dirt and grease will not collect in it. The finger plate shown in Fig. 280 provides a little more room for the fingers than the one shown in Fig. 278. It consists of a hinged plate with a small spring attached to keep the plate in position. A piece of leather or cloth is tacked over the bottom of the hole to prevent dirt collecting in it.

FOOT RESTS

In the majority of touring cars, and in some roadsters, a foot rest is provided for the convenience of the passengers. In Fig. 281 is shown a conventional design of foot rest which has a nickel-plated rod and nickel-plated or enameled end brackets. The rest shown in Fig. 282 is made from two end brackets and a center board, covered with a suitable material. A rest of this type is very comfortable, as it gives more surface to support the feet. There are a great variety of designs and types of foot rests. The design and location is largely a matter of individual taste. The two types illustrated are representative of the styles most in use.

ROBE RAILS

A robe rail is provided on almost all touring cars, and it is generally attached to the back of the front seat. The types in use can be divided into three styles: First, the solid; second, the folding, and, third, the flexible. In Fig. 283 is shown a solid type robe rail located in the position that is generally acceptable. The folding rail is shown in Fig. 284. This rail folds down against the back of the seat when not in use as shown in Fig. 285. The flexible rail, shown in Fig. 286, is favored by some because it can be detached at one end and the robe set on it before it is buckled again. The details of these rails vary, but a more elaborate description is unnecessary.

VENTILATORS

Ventilators have been used on and off by car manufacturers ever since the four-door body was made. While it is a great convenience for the passengers in hot weather, it is difficult to attach to a modern car without breaking up the lines and surfaces. The ventilating type of windshield has taken care of the ventilating conditions to some extent, but it is not as effective as a regular ventilator fitted on to the top or sides of the shroud. In Figs. 287 and 288 are shown different types of ventilators that are used. A type of ventilator that is designed to attach to the shroud is shown in Fig. 287. It is opened and closed by moving a circular plate around so that the openings in it register with the opening in the other plate of the ventilator which it sets against. A fine meshed gauze screen is used to keep dust and water out. The small screen circular cover is set to catch the air. The ventilator shown in Figs. 289, 290, 291 is of rectangular shape; this is hinged so that it may be pulled out as required. When it is pulled out it will catch the air; the fine gauze metal covering keeps the rain and dirt out. This type is more adaptable to fit on top of the shroud, but it can be used on the sides.

In Fig. 293 is shown an auxiliary device for ventilating. This is an attachment that can be put on the door post to hold the doors open and so permit the current of air created by the car, when it is moving, to enter the compartment. This device is especially effective when the doors are hinged at the back, but if the door

FIG. 289
Top cowl ventilator (G. W. Henvis, Philadelphia)

FIG. 290
Side cowl ventilator (G. W Henvis, Philadelphia)

FIG. 291
Top cowl ventilator in place (G. W. Henvis, Philadelphia)

FIG. 292
Roof ventilator for limousines and other enclosed cars (Waterhouse & Freeman Mfg Co., Newark, N. J.)

is hinged at the front the ventilating windshield has to be relied upon to force the air down into the compartment and out at the doors. The device illustrated consists of a bar with two hinged ball ends. These ends engage in two plates that are attached to the door post and door frame of the car respectively. These post plates have keyhole slots in them that permit the ball end to be readily slipped into position. The enlarged view in the sketch shows more clearly how this bar sets in position, and the section shows the construction of one end.

A later development in ventilation is shown in Figs. 289 and 290. These constructions are very practical and simply made, and easily operated. They permit a liberal sized opening to be used. The top cowl ventilator shown in Fig. 289 is operated by means of the projecting lever, while the ventilator which goes in the side of the cowl, Fig. 290, is operated by a knob that is pushed in to operate. It is made so it can be stopped at any point, allowing whatever degree of ventilation desired. Fig. 291 shows the upper ventilator assembled to the body. In order to provide proper ventilation for a closed car, one or more ventilators are put on the

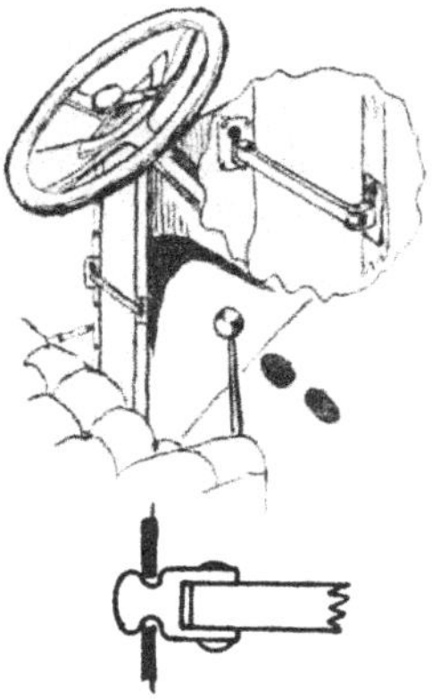

FIG. 293
Device placed on door to hold it slightly open for ventilation

roof. When one is used it is put in the rear and when two are used one goes at the front. When two are used it is generally to take care of the double compartment body, such as a sedan with the front seat divided off. A roof ventilator is shown in Fig. 292.

FIG. 294
Ternstedt windshield wiper
(Ternstedt Mfg. Co., Detroit)

A windshield wiper or cleaner is becoming a part of the car's regular equipment, and is certainly a very useful article in rain storms. There are a large number of different kinds of windshield wipers, some typical designs being shown in Figs. 294, 295 and 296. Automatic wipers are shown in Figs. 295 and 296. The first is operated by the engine suction and will swing back and forth when the valve is opened, and the second, Fig. 296, is electrically operated.

Rear vision mirrors are becoming standard equipment on closed cars and are used quite extensively on open cars. Like the majority of automobile equipment there are a large number of companies making them. A typical rear vision mirror is illustrated in Fig. 297.

Closed cars, such as town cars, limousines and landaulets, are equipped with a speaking tube or telephone. Modern electrical development has provided very efficient equipment, one of which is known as the "autophone." The complete equipment for a car is shown in Fig. 310. The mouth piece is made in any design to match the exterior hardware finish. (See page 146).

FIG. 295
Mayo-Skinner Windshield Wiper

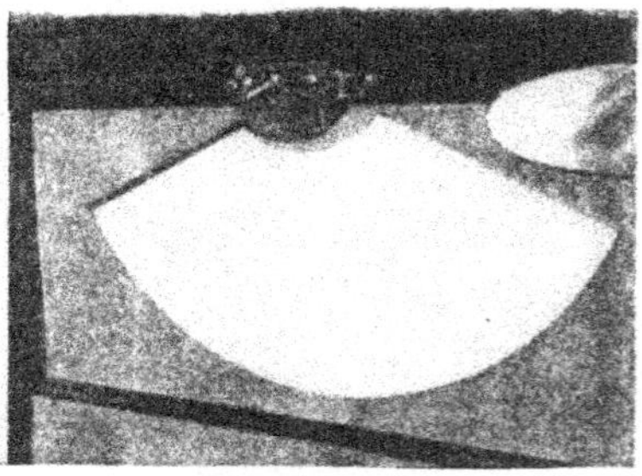

FIG. 296
Trico Windshield Wiper

FIG. 297
"Pioneer" Auto Mirror for open and enclosed cars. (Brewer-Titchener Corp., Cortland, N. Y.)

SUNSHADES FOR OPEN AND CLOSED CARS

Sunshades, or visors, for automobiles are becoming very popular. While they have been in use more or less for several years, it is only within the last twelve months that they have made such rapid strides in gaining popularity. As far as closed cars are concerned the visor can be classed as standard and necessary equipment. Very few closed cars are marketed nowadays which do not have a sunshade as regular equipment. On the closed cars a three-piece windshield has been practically standard equipment for several years excepting on some of the cheaper cars. This shield was generally constructed with the third glass hinged so it could be raised or lowered in a certain range similar to that shown in Fig. 298. The main idea was that it gave extra protection to the inner windshield in bad weather and if it were necessary to drive with the inner glass open the outer one could be set so it would prevent the rain from coming in. These things it would accomplish to a slight degree and when it became dusty it would help to protect the passengers from the full glare of the sun when driving against it.

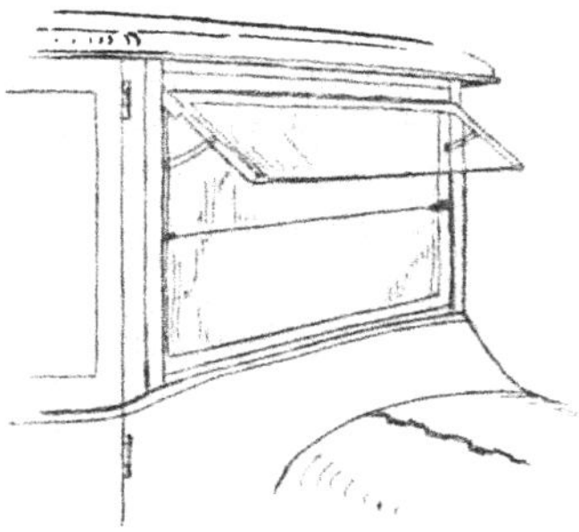

Fig. 298
A three-piece windshield equipment has been standard for several years on all but the cheaper cars

The average car owner is accustomed to driving on trips which take him in all directions; for instance, if he took a trip north or south he would not get very much of the direct rays of the sun in his face, only when he was angling or riding east or west which would only be for a short time if his main direction was north or south. Should his trip take him east in the morning and west in the evening he would find the direct rays of the sun very trying. There are many people who have fairly long trips to and from business who go east to business in the morning and west back again and they would always be driving against the rays of the sun. Numbers of people have gotten along without a sunshade, but if they ever installed one they would never be without one again. Of course no shades are made one hundred per cent. perfect or efficient, because the undulation of the road makes the line of sight vary in a manner similar to the light from the headlights which throw their lights into the sky going up hill and into the road going down hill. In sketches Figs. 299 and 300 are shown how the line of sight varies.

It may well be asked why visors are not universally adopted for the open car as well as for the closed ones. The answer is that the accessory houses are selling a lot of them. The average open car has quite an extension of the top over the windshield so it is not so necessary to have a visor as this extension gives a fair amount of sun protection. This condition is shown in Fig. 301. If more protection is desired it is very easy to buy a visor and fixtures, as there are a large number of new designs of suitable shades.

The sunshade for a closed car can be made in a large number of different ways, several of which are shown in sketches No. 302 to No. 306. In Fig. 302 is

Fig. 299
Showing angle of sun's rays on windshield when the car is going up hill.

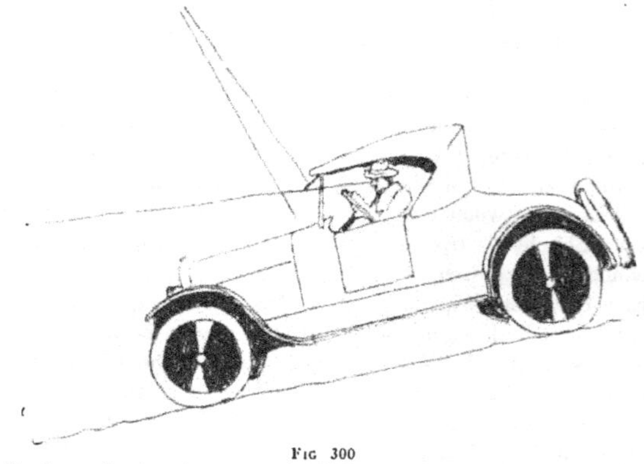

Fig 300
Showing angle of sun's rays on windshield when the car is going down hill.

shown a semi-permanent style of shade. It consists of two brackets with a fabric cover stretched over them and secured by curtain fasteners. This will allow the driver to remove the fabric shade or else roll it up so it will not be in the way. The type shown in Fig. 303 is permanent; the cloth can not be easily removed and adjustments are impossible.

In Fig. 304 is shown one that can be folded out of the way when not required. The body is of cloth and it works on a roller similar to a house spring-shade roller, which will roll up when jerked slightly to release the catch. The shade shown in Fig. 305 is made of metal and is adjustable to only a small amount as shown. The

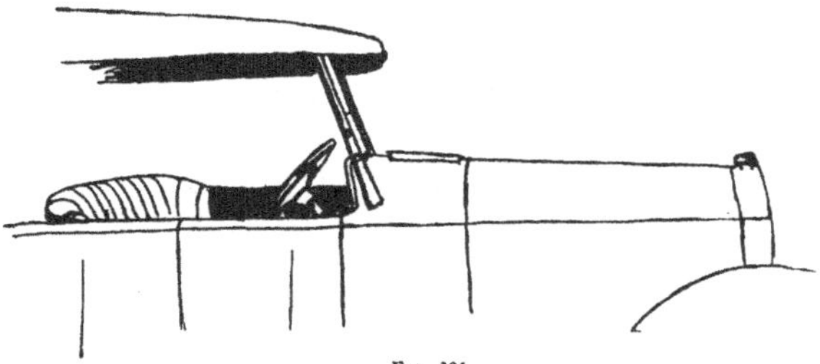

FIG. 301
When top of car extends out over the windshield it is not so essential to have a sunshade.

holes in the bracket permit about 20 degrees of movement which will take care of all ordinary requirements. The shade in Fig. 306 is made of parts similar to those used with the first glass-type windshield, but instead of using a glass a board is inserted. The brackets on the side allow it to be adjusted as desired. The same

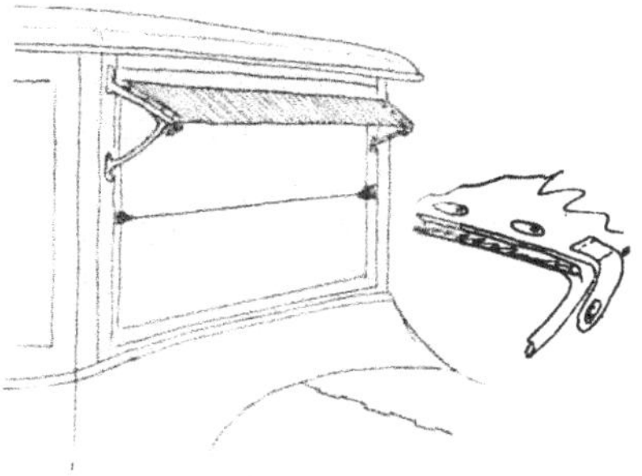

FIG. 302
In this sketch is shown a semi-permanent type of sunshade. The fabric may be removed or rolled up

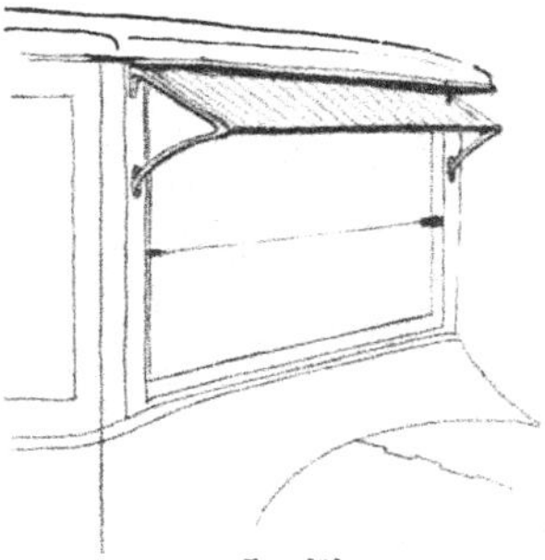

FIG. 303
This type of sunshade is permanent and no adjustments can be made.

general construction can be used as shown in Fig. 306 except that instead of wood a sheet of glass is employed. This glass can be either blue, green, opaque, or brown; and it can be made of two colors, the upper portion opaque and the lower brown, blue, green or any other color desired. This type of shade is quite attractive in appearance, but it is not very much more efficient than a cloth one, for while it is possible to see through them it is also possible for the sun's rays to penetrate slightly.

The average open car does not have sufficient of the top projecting over the windshield to protect the driver from the direct rays of the sun, consequently

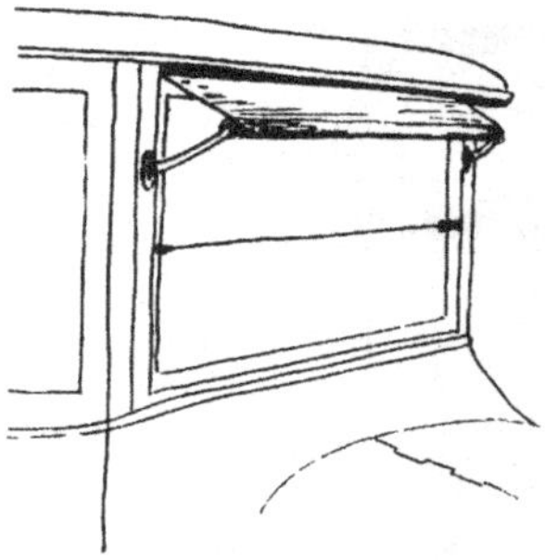

FIG. 304
This sunshade has hinged brackets and the cloth is attached to a spring roller. A slight jerk releases the catch and the shade folds up when not needed.

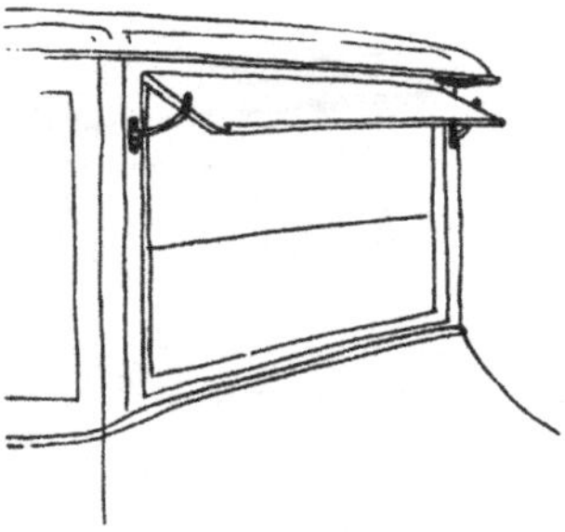

FIG. 305
This sunshade is made entirely of metal and is adjustable to a small degree.

FIG. 306
This sunshade takes the place of the third member of a three-piece windshield but a board is used instead of glass.

there are quite a number of companies manufacturing sunshades suitable for use on a touring car, which make an assembly similar to that shown in Fig. 307. This is made from a metal frame and side supporting brackets. The under side of the frame can be covered with green imitation leather, as it makes a much better appearance than the black.

Any top builder would find it profitable to be able to make up a set of sunshades. Any new top he makes should include a shade of some kind, and it will be found that the customer will appreciate this added touch to the work and be quite willing to pay for it. If the top is not collapsible the sunshade brackets can

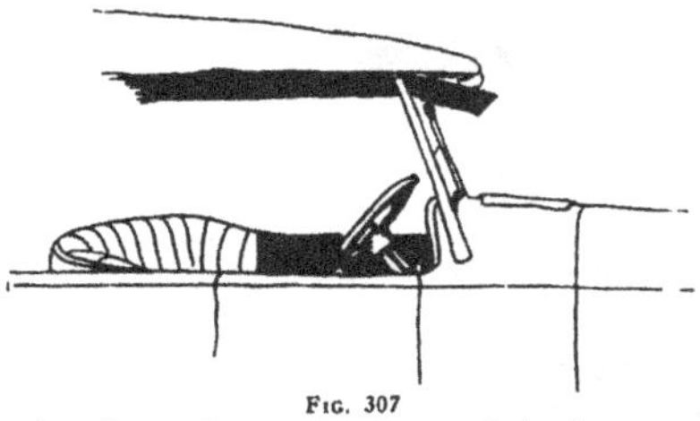

FIG. 307
There are quite a number of companies manufacturing sun shades for use on touring cars which make an assembly similar to the one sketched here.

FIG. 308
When a sunshade is fitted to a collapsible top, the shade is fastened to the windshield posts.

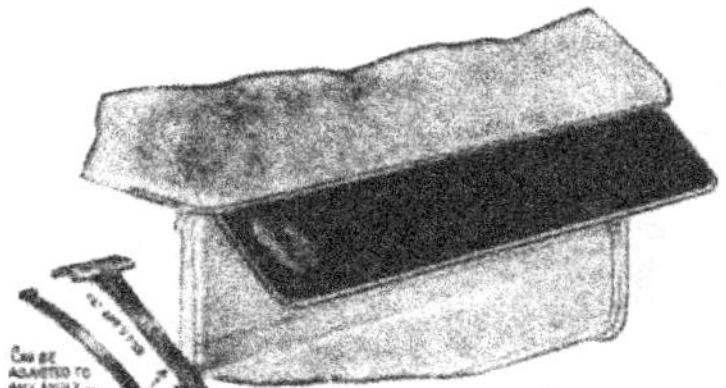

FIG. 308 A
"Monarch" Windshield Visor (Monarch Carriage Goods Co., Cincinnati)

be fastened on to the top bow. Should it be fitted to a collapsible top the shade will have to be attached to the windshield posts. Sketch Fig. 308 shows one way that this can be worked out at a reasonable cost. Fig. 308 A is a sunshade of a popular style recently put on the market.

FIG. 309
• Mechanical window regulator.
(Jos. N. Smith & Co., Detroit)

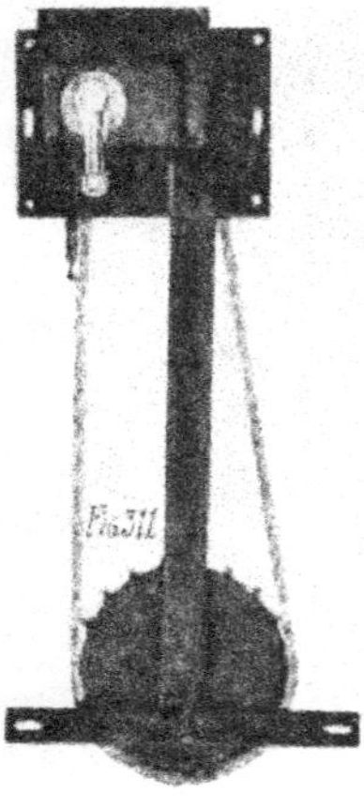

FIG. 311
"Perfect" window regulator
(Perfect Window Regulator Co., New York)

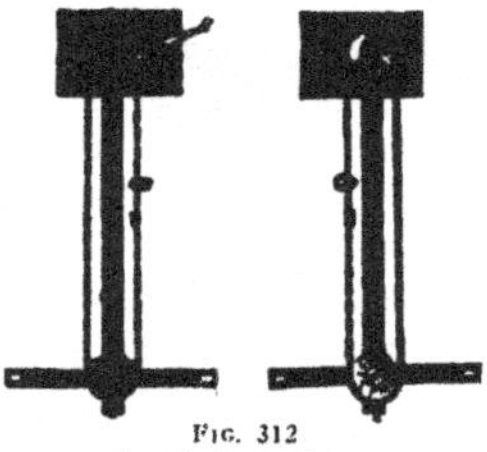

FIG. 312
Standard regulator

FIG. 315
Center partition assembly

FIG. 314
An offset regulator

FIG. 316
Rear quarter assembly

The window regulators in Figs. 312 to 316 are made by Ternstedt Mfg. Co., Detroit

WINDOW REGULATORS

This chapter on "general hardware" would not be complete without mentioning window regulators. They constitute, in fact, a very important item. It used to be "good enough" to have the windows operated by straps, and while the straps are still used quite extensively for the rear quarter windows, very few cars are built without a mechanical lift in the doors. There are several different types of window regulators, but they are all patented constructions, consequently they are

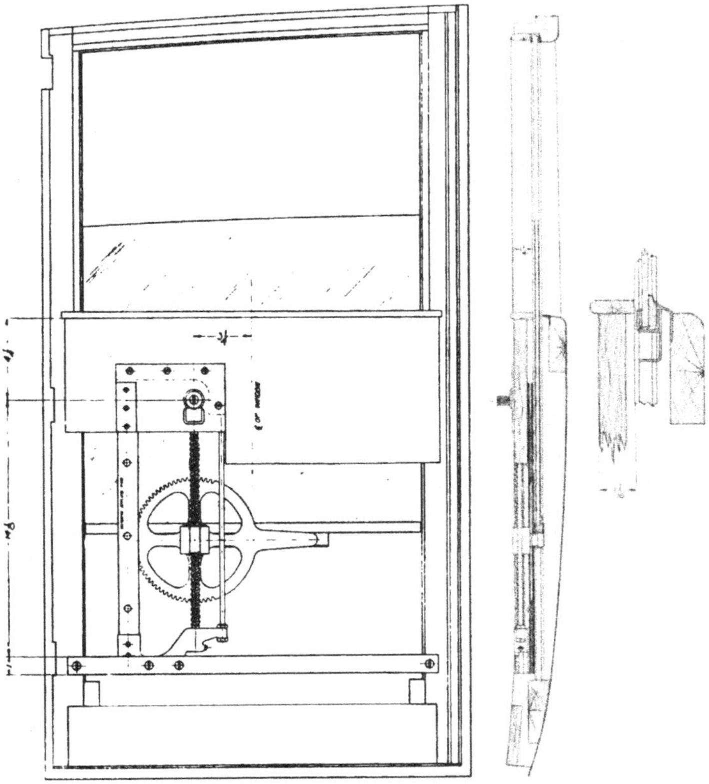

FIG. 309 A
Reproduction of factory blue print, showing construction and method of installing the Jos. N. Smith & Co. window regulator

of different design. Among others, there are the Perfection, Ternstedt, Smith, Dura and Perfect. These are made in different lengths and assemblies, so that they can be adapted to any conditions. When their installation is contemplated detail blue prints are furnished for the body engineer's information. The Perfect regulator is shown in Fig. 311 and in Fig. 312 is shown a standard Ternstedt

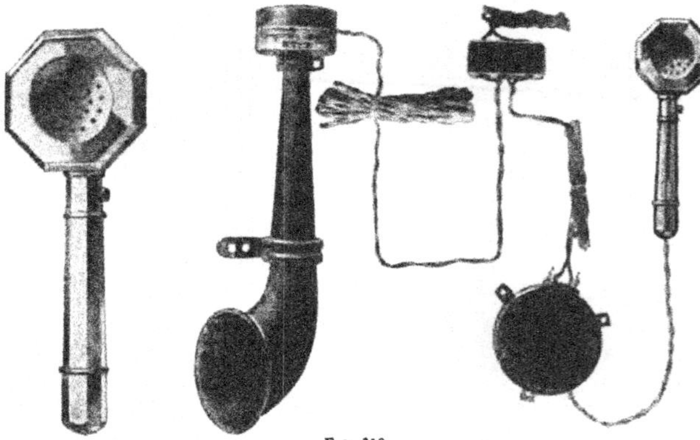

FIG. 310
Stentor Autophone
(Klaxon Company, Newark, N. J.)

regulator. An offset design is shown in Fig. 314 and Fig. 315 is a center partition assembly. Fig. 316 shows a rear quarter assembly.

The Dura regulator, in its latest development, operates by turning or rotating a handle, similar to the other types. Fig. 309 is the new window regulator introduced by Joseph N. Smith & Co. during the winter of 1922. A factory blueprint of same is shown in Fig. 309A.

CHAPTER XV

FENDERS

The rear fenders of the latest designs have a small wing fitted in the rear, this closes up the small opening that is left between the back of the body and the rear of the fender. While it is only a small wing it prevents a great amount of dust and mud from being drawn up the back of the car by the vacuum caused by the car's displacement in the air. There are a number of different ways to design a rear fender wing, but the design shown in Fig. 318 is one of the best. It will be noted that it extends well down to the end of the fender at the rear and then runs along under the body and bolts against the frame. It is riveted or bolted on to the fender and fastened by screws or bolts to the bottom of the body.

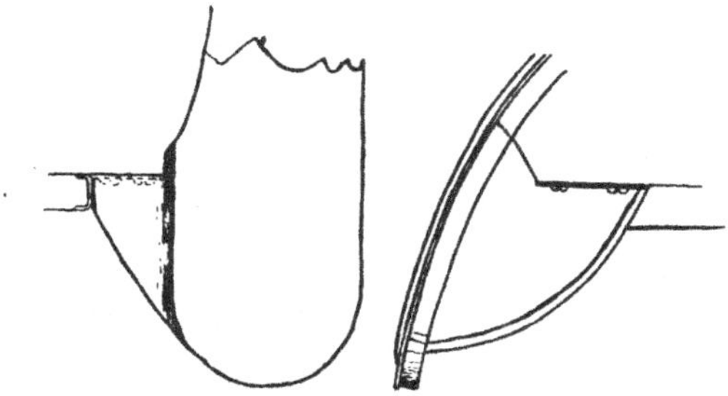

FIG. 318
One of the best ways to design a rear fender wing

So far the touring car fenders have been described, and so the roadster fenders must be considered next. When designing the roadster fenders it is general practice to use the same front fenders that are used on the touring car. This avoids making two sets of front fenders, and is a decided economical factor in manufacturing automobiles. As it has been considered preferred practice to use the touring car front fender only, the rear fender will be described. There are two methods used to attach the rear fender to the car, and this will be used to mark the difference in the styles.

The first design, and the most popular one, is to make the fenders along the same lines as the front ones, that is, to have a formed wing inside and carry the fender on suitable irons that attach to the frame. This is shown in Fig. 319. In the second design the fender attaches to the body somewhat in the same way that the rear fender on the touring car does. This design is shown in Fig. 320. This design is easier to make and to assemble to the car, consequently it is cheaper. It

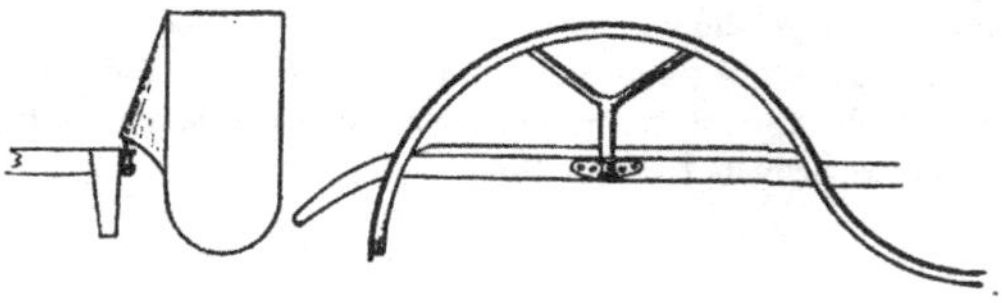

Fig. 319
Rear fender with formed wing inside, carrying the fender on suitable irons that attach to the frame

does not look as well as the first design and also the body must be designed to suit it. When the body is not shaped to fit this second design it is very difficult to make it look well.

While the general practice in roadster fenders has been outlined in the foregoing, the designers, in their endeavors to make their roadster distinctive, have made a great variety of different fenders. These are interesting for reference, and for those who want to spend the money to get distinctive results, so in Figs. 321, 322, and 323 are shown various types that are representatives of their class.

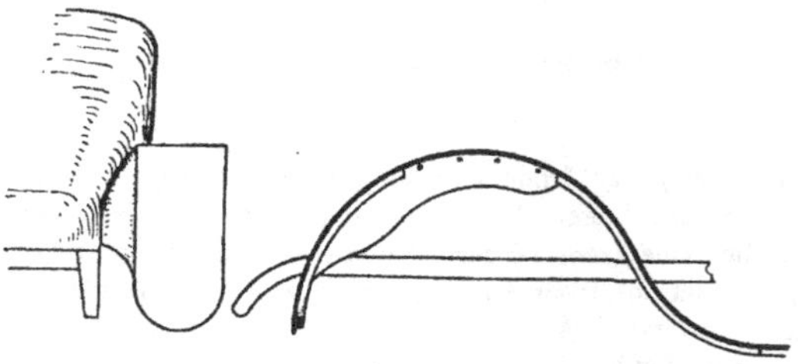

Fig. 320
Rear fender attached to the body, as is the practice with many builders

The methods used to attach the fenders to the car have gone through some interesting stages in their development. The front and rear fenders were supported on forging irons that were attached to the frame. The front fender irons were very heavy and had long arms that extended to the front and rear of the fender and riveted to it; the iron was attached to the frame by bolts or set into a taper socket. The rear fender iron also had extended arms that fitted to the fender in a similar way to the front ones.

The front fender irons were then simplified and metal plates used instead of arms to attach the iron to the fender, and next the metal plates and forged irons were eliminated and a pressed steel iron used. The pressed steel iron is now used on a large number of cars. It is much easier to make than forged iron and is lighter. The front fender iron is now very largely used to carry the headlights and so saves the weight of the forged steel headlight supports. In Figs. 324, 325 and 326 are shown three different phases in the developments of the front fender irons, these illustrations show the basic designs.

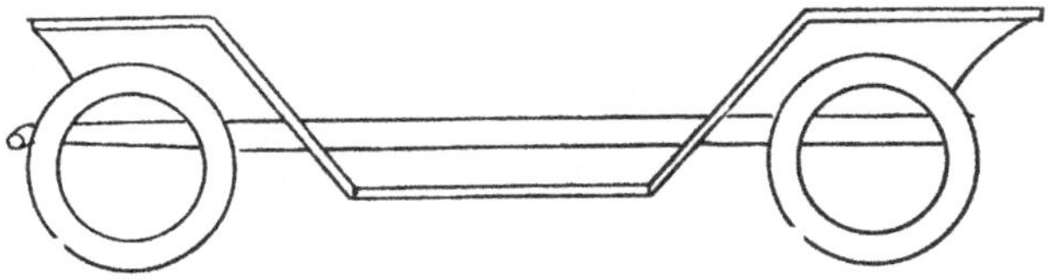

FIG. 321

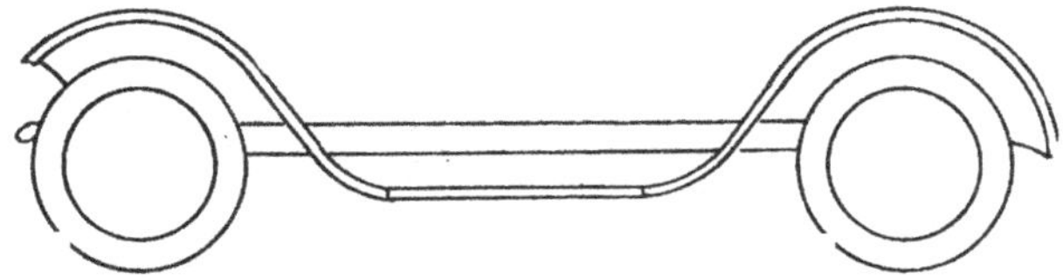

FIG. 322

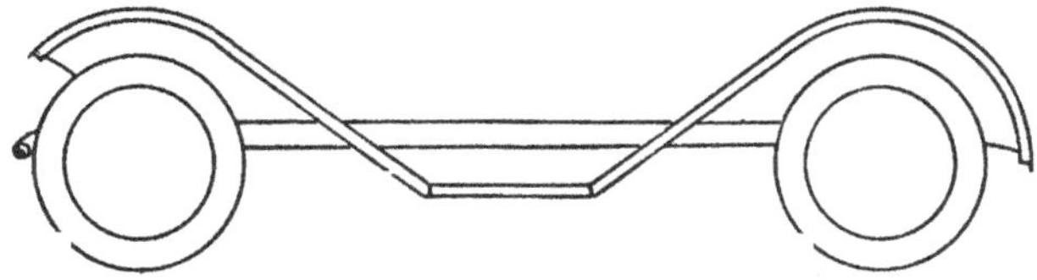

FIG. 323

Figs. 321, 322 and 323 are interesting for reference, as they are types representative of distinctive treatments of the fender problem

In Fig. 324 it will be noted that the fender wing is short and that it is secured to the irons by small strips of steel. These are bent over the iron and riveted to the wing. This clamps the wing up tight to the iron and holds it from rattling. The fender shown in Fig. 325 has a larger and longer wing. This extends down to the frame and is secured to the frame by a bracket that bolts on to the frame and clamps the wing tight against it. By clamping the wing down in this way a rear iron can be dispensed with. In Fig. 326 is shown the same style of fender having

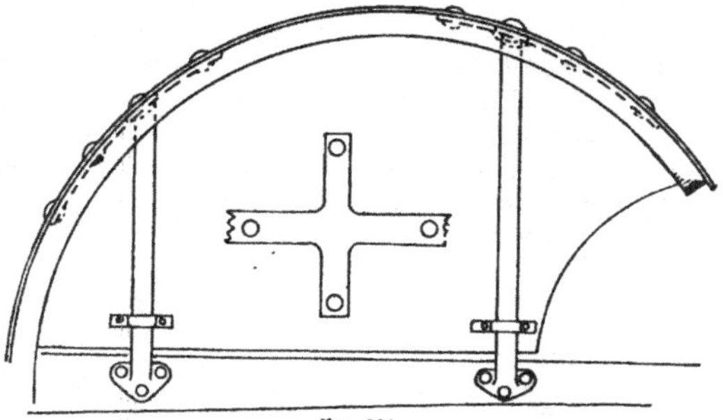

Fig. 324
Here the fender wing is short and is secured to the irons by small straps of steel

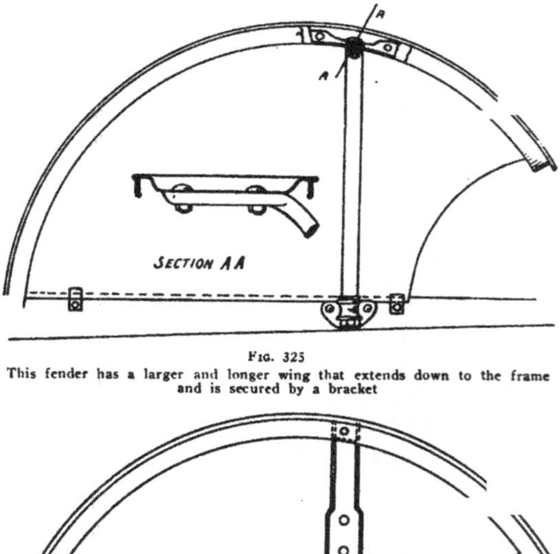

Fig. 325
This fender has a larger and longer wing that extends down to the frame and is secured by a bracket

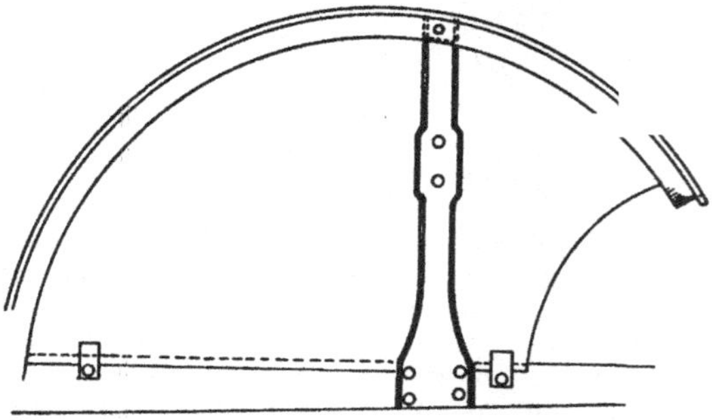

Fig. 326
Same style of fender as Fig. 325, but attaches direct to frame by four bolts, without brackets

the wing clamped to the frame. The iron, which is of pressed steel, attaches to the frame by four bolts. The fender is fastened to the iron at the top only by two rivets. In Fig. 327 is shown a side view of the fender iron, and the dotted lines show how the headlight is attached to it by bolts. The rear fender iron shown in Fig. 324 has been eliminated and the fender is fastened on by cap screws as shown

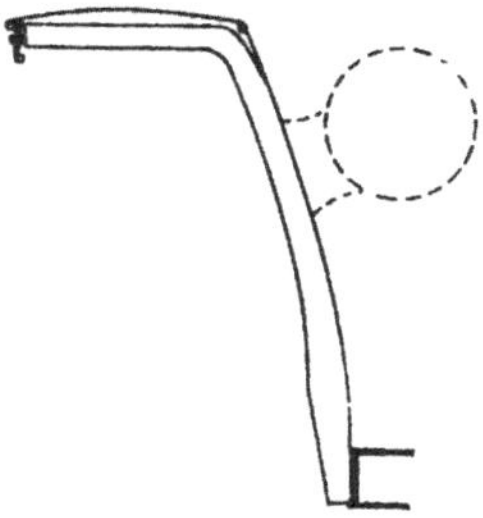

FIG. 327
Side view of fender iron, the dotted lines showing how the headlight is attached to it by bolts

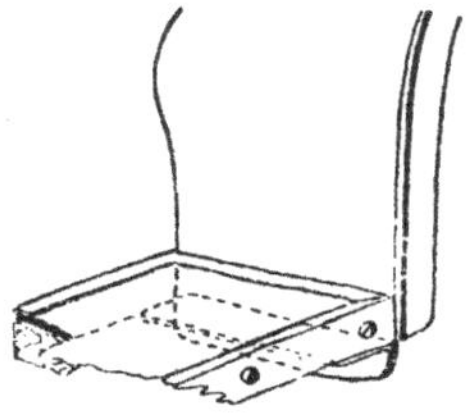

FIG. 329
This rear fender has the bottom end built forward under the running board, two carriage bolts holding the fender in place.

in Fig. 326. The roadster rear fender iron went through the same development as the front fender iron. In Fig. 326 is shown how a pressed steel iron is made to carry the modern front fender.

The running board fills the space between the front and rear fenders and makes a suitable stepping place for passengers when getting in and out of the car.

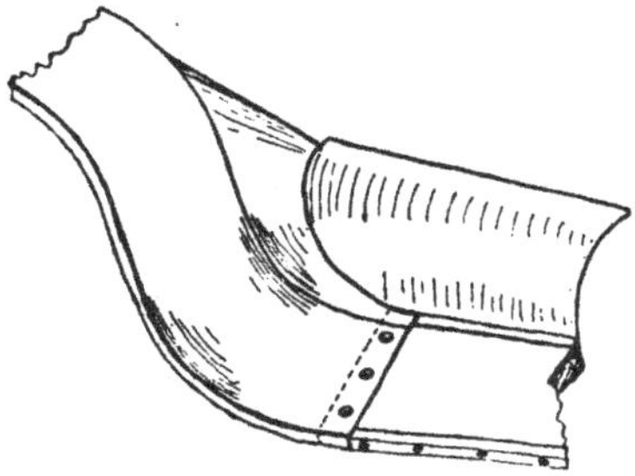

FIG. 328
Front fender set on top of the running board and held in place by three carriage bolts.

It is standard practice to extend this board from fender to fender without a break, supporting it by suitable irons fastened to the car frame. The fenders attach to the front and rear of this board, and as there are a variety of ways to do this, two methods have been chosen to show the best means to accomplish it. In Fig. 328 is shown the front fender setting on top of the running board and held in place by three carriage bolts. The rear fender shown in Fig. 329 has the bottom end built forward under the running board, two carriage bolts set down under the linoleum on the board and hold the fender in place. The section shown in Fig. 330 explains this. A finished plate fastens over the tops of the bolts by two screws and covers the holes the bolts make in the linoleum. In Fig. 331 is shown the front fender having its end bent down and an extension plate welded to it, an angle is bolted to the running board and against this the fender is held by bolts going through it

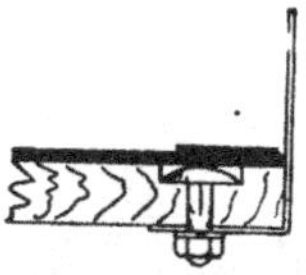

FIG. 330
Section showing detail of Fig. 329

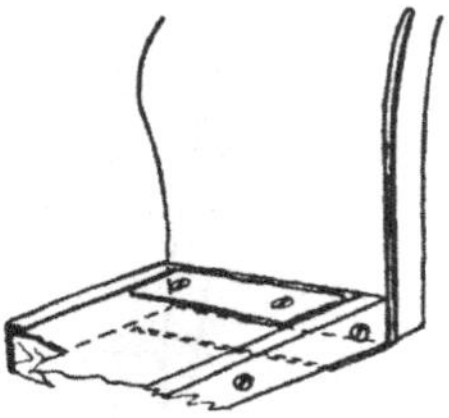

FIG. 333
A method of fastening rear fender to the running board.

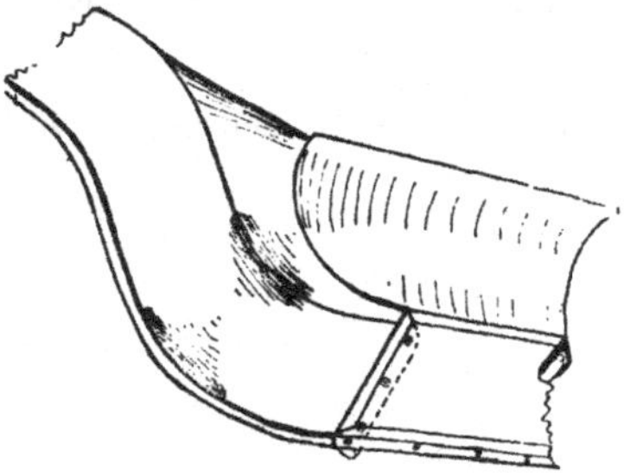

FIG. 331
Front fender having its end bent down and an extension plate welded to it.

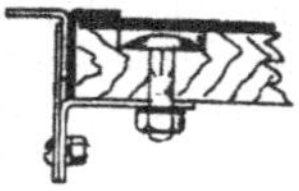

FIG. 332
Section showing detail of Fig. 331

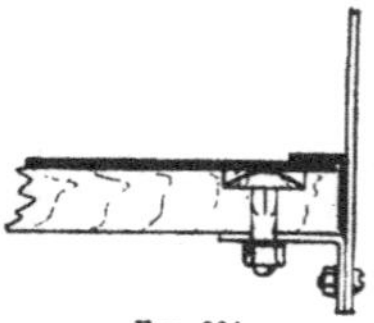

FIG. 334
Section showing detail of Fig. 333

into the fender extension, the sketch in Fig. 332 shows this very clearly. The rear fender fastens to the running board in a similar manner, and an angle is bolted to the board, and the fender has an extension that matches the angle on the board. Bolts go through them both, and securely fasten the fenders at the bottom. This is shown in Figs. 333 and 334. The first method described is a very satisfactory one for economical manufacture, as the parts are easily made and quickly assembled. The second method is more complicated, but permits of a uniform binding being attached to the board on all four of its edges. This makes a better looking, but more expensive, finish than the first method.

The side aprons, or running board skirts, are metal covers that interpose between the lower edge of the body and the running board, and protect the board from mud that would otherwise splash up. In the earlier cars the space between the frame and the running board was left open. This allowed the mud to splash

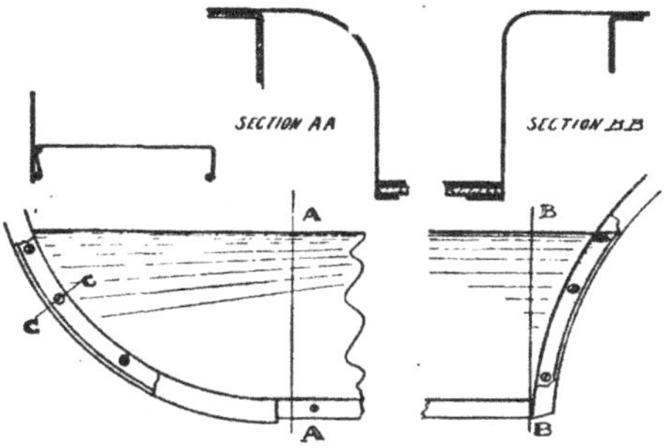

FIG. 335
A side apron that is a good example of current designing practice, showing method of attachment.

up the boards. To prevent this a metal cover was provided and this has now become a standard American practice. The apron fills up the space under the body and between the front and rear fender, and makes a clean and smooth finish, instead of leaving an unsightly gap exposing the springs and running board supports between the bottom of the body and the running board.

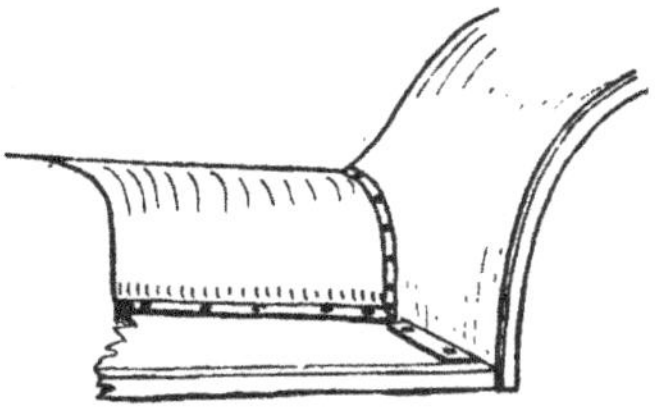

FIG. 336
The older method was to bolt the side apron on to the running board. A half-oval moulding was welded to the side apron and this was countersunk for French head bolts.

The side apron is made from light sheet metal which is generally 22 U. S. gauge. It is so formed that the top edge rests on the car frame under the body, and the lower edge extends down to or past the running board. The front end of the apron has a larger exposed surface than the rear, owing to the fact that the body gets wider at the back. In order to make the car look balanced, the apron is pulled in closer to the frame at the front end to give it an appearance of being about the same size as the rear. In order to do this the running board has to be made wider at the front. The side apron shown in Fig. 335 is a good example of the prevailing design and shows in the section how it sets under the running board and on top of the frame, while the ends project inside of the fender and is bolted to the flanges on them.

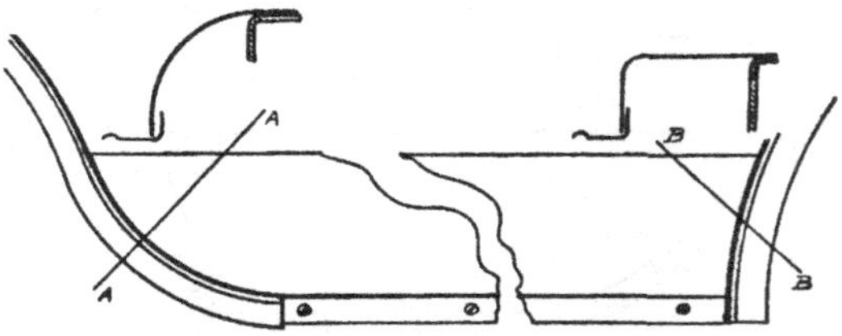

FIG. 337
A cheap method of assembling a side apron to a car, which does not protect the running boards from the mud which works in between them and the fender.

There are a large number of ways to fasten the fenders to the side apron, and the method shown in Fig. 335 is one of the best. The old method used was to bolt the side apron on to the running board and to the side apron a half oval moulding was welded, and this was countersunk for French head bolts as shown in Fig. 336.

A very cheap method of assembling a side apron to a car is to have the ends cut to match the contour of the fenders, allowing a slight clearance. The fenders

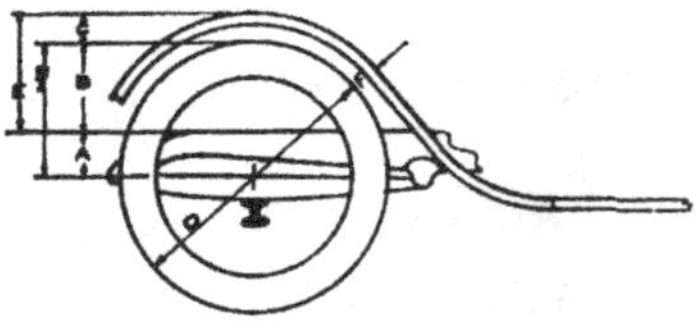

FIG. 338
Front end of the frame with wheel, fender and hood in their relative positions.

have flanges turned on them that project under the side apron. No screws or bolts are used to fasten the apron to the fender. The fender flanges prevent the gap from being noticeable, but do not protect the running board from the mud which will work in between them and on to the fender. In Fig. 337 is shown this style of side apron.

The different styles of fenders have been described, starting from the earlier types and tracing the progress up to the present practice. In studying a subject the past history is always interesting, and in some cases will save a designer a lot of time when he can start his work from the modern designs, for, knowing what has gone before, he will be able to avoid the errors of the earlier designs.

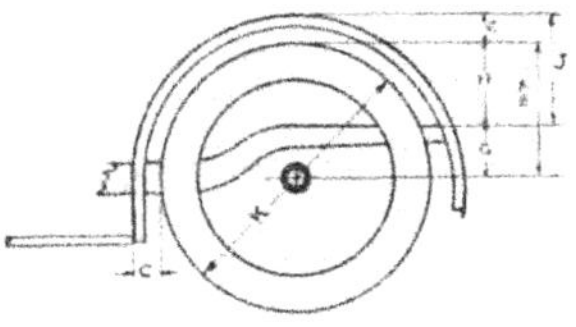

FIG. 339
Rear touring car fender showing its relation to adjacent parts of the car.

In starting a design the basic principles should be considered first. When these are known the conditions that are to be determined in relation to the design that is planned are quickly located. In the diagram shown in Fig. 338 is outlined the front end of the frame with wheel, fender and hood. "A" stands for the distance from center of the wheel to the top of the frame when the spring is compressed or loaded so that the bumper strikes the frame. "B" is the distance from the top of the frame to the top of the tire, and "C" the clearance that is required between the top of the tire and the fender for non-skid chains; "E" will be the dis-

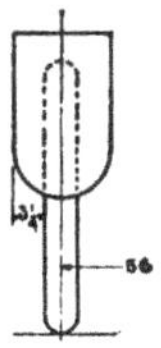

FIG. 340
How the front fender looks with a 3½ inch tire and an allowance of 3¼ inches from the outside of tire to outside of fender.

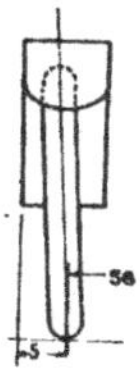

FIG. 341
Here it is seen that the fender now has ample overhang, as much as can be allowed to meet the conditions explained in the text.

tance to locate the top of the fender above the frame. Where D equals radius of the wheel, E equals D plus C minus A. C is generally made between 1½ to 2 inches. The front edge of the fender should line up with the front of the tire, and the rear end should line up with the back of the engine bonnet. A point that must be watched is marked F. This distance should exceed the distance C just enough to make the fender appear to have an equal amount of clearance all around the wheel from the front back to a point where the fender starts to leave the wheel line and bend towards the board.

The rear touring car fender is shown in the diagram Fig. 339. G equals the distance from the center of the wheel to the top of the frame when the spring is compressed so that the bumper on the axle strikes the frame. "H" is the distance from the top of the frame to the top of the wheel, and "C" is the distance required for chain clearance. The distance J is the height to locate the top of the fender above the frame. The radius of the wheel is K; then $J=K+C-G$. The fender starts at the running board and generally has a clearance C equal to 2 to 3 inches, similar to the top. As this point only varies in a vertical plane, it is located as close as possible.

The rear fender follows the same principles as the front; also the roadster rear fender uses the same rules to start the designs from. The widths of the front and rear fenders are generally made the same and the over-all width is practically standard. The track or tread has been standardized in America to 56 inches. Previous to this a 60-inch track was used in the South. The over-all width depends on the size of the tire. Taking a 3½-inch tire, the width of the car over the tires will be $56+3\frac{1}{2}=59\frac{1}{2}$, while with a 5-inch tire it will be $56+5=61$, a difference of 1½ inches. This is not very much. It is average practice to allow 3¼ inches from the side of the tire to the outside of the fender. This will give an over-all width of 66 inches on a 3½-inch tired car, and 67½ inches on a 5-inch tired car.

In Fig. 340 is illustrated how the front fender looks with a 3½-inch tire and an allowance of 3¼ inches from the outside of the tire to the outside of the fender. These measurements are obtained by measuring at the track of the wheel. By taking measurements at the track the main clearance is obtained, owing to the fact that the front wheels toe out at the top while the rear wheels do not do this.

By inspecting the sketch in Fig. 341, of the rear fender and rear wheel, it is seen that the fender has ample overhang to it now. Comparing it to the front fender in Fig. 340 it is seen that the tilt of the wheel cuts down this projecting overhang, but this is as much as can be allowed to meet the conditions explained above. These points that have been enumerated are the basic principles of fender design, the shape and lines are worked out from these determining points in accordance with the designer's ideas.

DEFECTS OF PRESENT FENDER DESIGN

The developments of the fenders have been traced up to the present styles. While these are effective and blend with the beautiful lines of the modern car, they leave much to be desired to be called "ideal." They protect the car and its occupants well under certain conditions only. On bad, muddy roads the car is quickly covered with mud, and even on fair roads the car is soon bespattered, if only moderate speeds are indulged in. After a rainstorm in the city, pedestrians are spattered with mud if the traffic is at all plentiful. A small collection of water, that is hit by an automobile traveling at medium speed, is splashed for a good distance in all directions. In towns that have narrow streets, not only do the sidewalks get splashed with mud, but the store windows also are subjected to a liberal bath of mud. When it rains the mud appears like magic, even on the cleanest appearing streets in the cities. Of course this condition is very much aggravated on unpaved streets.

A few years ago the officials in Paris, France, offered prizes to those who could invent fenders or attachments for them that would prevent splashing at certain speeds. The results of their trials were not satisfactory; the only device that was anywhere near efficient was a brace-like attachment that swept along the ground at the side of the wheels. From the foregoing it is seen that here is a problem that has yet to be solved to make a fender that will protect the car, its occupants, and pedestrians as well.

The majority of automobile fenders are finished in black by enameling. This is the prevailing way of finishing the fenders and side aprons. It is both an economical method and gives a lasting finish. A fine, hard coat is obtained by baking and this gives a high gloss. A great advantage to the manufacturer is that the enamel is not easily marred in handling and in assembling on to the cars. A painted fender would require a lot of extra care taken so that it would not be marred in its progress through the factory. While the tendency in the higher priced cars is to finish the complete car in one color, this cannot be considered for the low and medium priced cars until a satisfactory process of color enameling is discovered.

CHAPTER XVI

WINDSHIELDS

The windshield is a comparatively modern fitting and it is only within the last few years that it has become a necessary part of the automobile. The first shields detracted very much from the general appearance of the car, as they did not harmonize with the rest of the construction, and to have a shield fitted to a car made it look freakish. Now that the windshield is standard equipment on virtually all cars, an automobile has an unfinished appearance when it does not have one on.

The development of the various types of windshields was a very rapid one, as the increased efficiency of the automobiles permitted higher speeds and some pro-

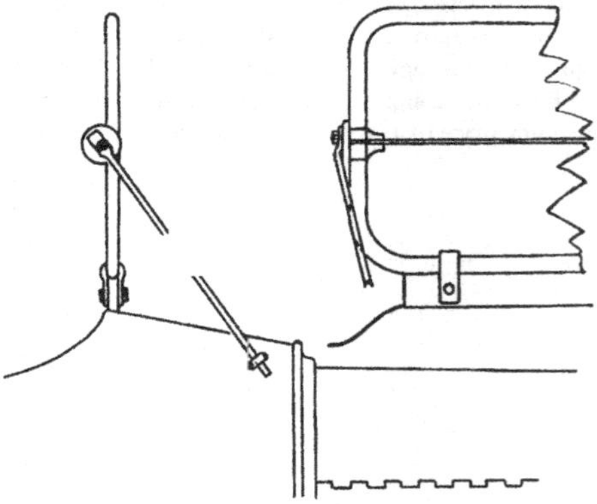

FIG. 342
One of the earliest types of windshields, fitted to brass or steel frames that clamped on the dash and were braced by long rods on the sides.

tection had to be provided against the dust and wind. When the designers found that the shield was necessary they soon discovered means for adapting a construction that would give a harmonizing appearance with the rest of the car.

The earliest type of shield, as shown in Fig. 342 consisted of suitable sized glasses fitted to brass or steel frames that clamped on to the dash of the car and were braced by a long rod on each side, either extending toward the front or toward the rear of the car. The shields were generally constructed in two sections, the upper one had a suitable hinge upon which it could be swung down, giving a clear view to the driver or passenger over the top of the lower shield. This type of shield is still used to a considerable extent, but the method of attaching it to the car is greatly improved.

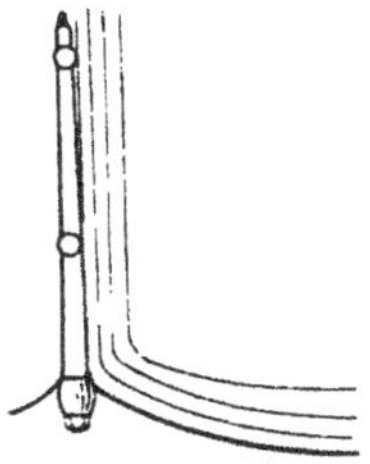

FIG. 343
The air impinges on the straight shield in this manner.

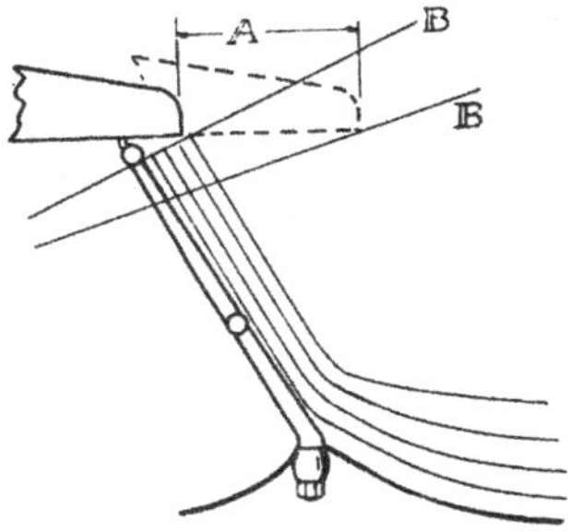

FIG. 344
The slanting shield reduces the necessary length of the top, resulting in better vision.

The desire to do away with the top straps that either extended to the front of the frame or attached to the cowl, resulted in the development of the side arm or "standard" type of shield. These standards were made high enough so that the

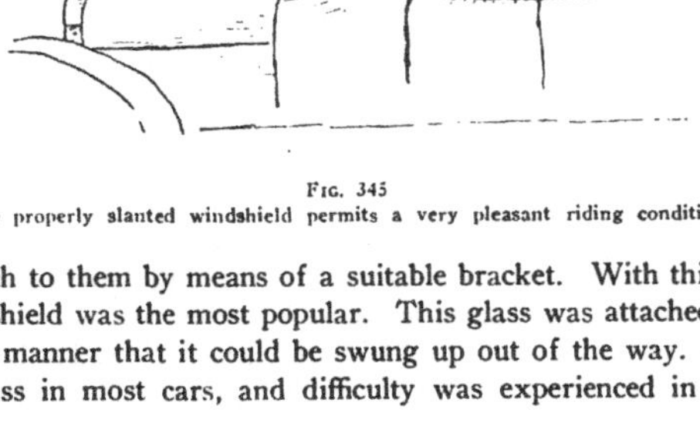

FIG. 345
The properly slanted windshield permits a very pleasant riding condition.

top could attach to them by means of a suitable bracket. With this development a single glass shield was the most popular. This glass was attached to the standards in such a manner that it could be swung up out of the way. This meant a large sized glass in most cars, and difficulty was experienced in holding it up when a free draft was desired. This sized glass was so heavy that it was frequently jarred down with the vibration of the car. The next design in development had two glasses, both of which hinged on the posts, so that they could be swung in any desired position. This type can be classed as the prevailing one for the present automobile. With a shield of this design a ventilating condition

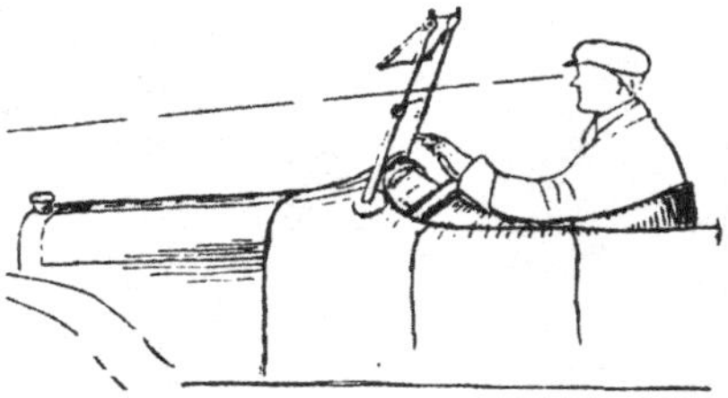

FIG. 346
The double shield can be closed in bad weather and in good weather opened at the top, while the lower portion deflects the air.

is obtained for the body by pulling the lower glass backward about two inches past the top of the shroud or bracket. This causes a current of air to be deflected downward into the compartment. For this reason this type of shield is termed the ventilating type; besides the ventilating feature there is a rain vision condition

FIG. 347
Under this condition there should be a ventilator in the top or an open window in the rear to avoid unpleasant air currents.

obtained by swinging the upper shield outward so that a gap of 2 inches to 3 inches is left between it and the lower one. This deflects the rain and permits the driver to look out between the two glasses, giving unobstructed vision.

For convenience of reference the different designs will be divided into five groups: First, the non-ventilating and folding upper glass; second, ventilated

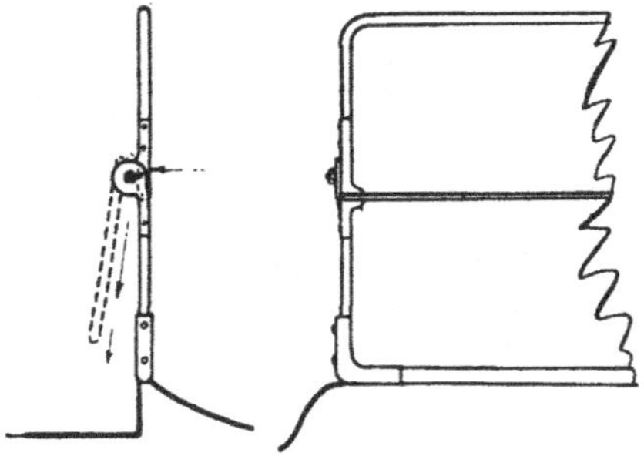

FIG. 348
Shield used on a low priced car that is made in great quantities.

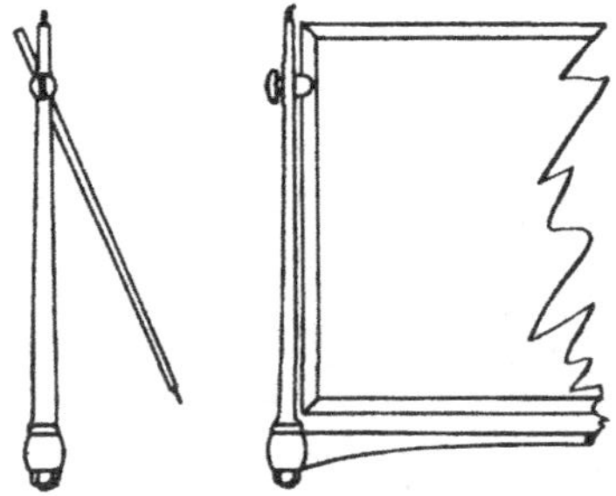

FIG. 350
This type is only adaptable to a short shield, as it is hard to keep the friction joint tight enough to hold it up for clear vision.

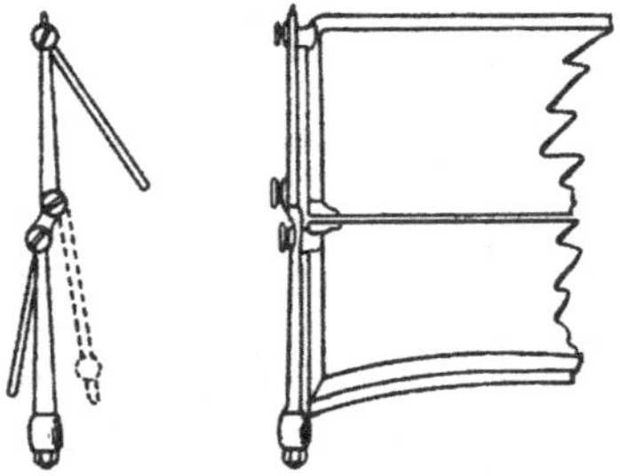

FIG. 349
There is a friction joint in each of the standards which allows the shield to fold as shown by dotted lines.

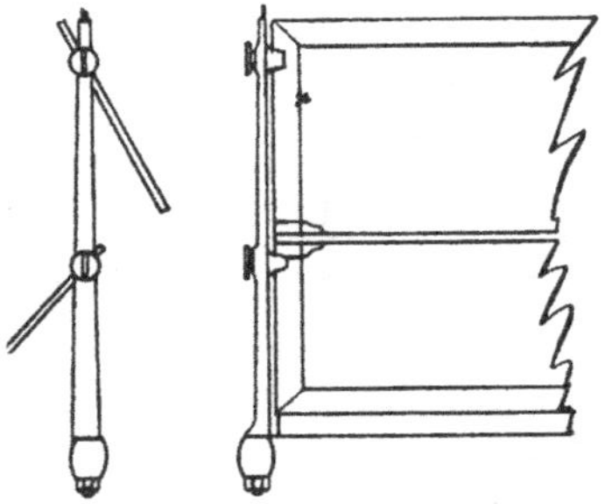

FIG. 351
This may be termed a "standard" shield. The drawing at the left shows the working range of the glass frames.

lower glass, top glass folds down; third, single glass; fourth, ventilating lower and rain vision upper; fifth, miscellaneous designs.

The modern windshield comes under class four, with variations in the details of its construction to suit the car which it has to be adapted to and also to meet the designer's ideas. The tendency is to set the shield on an angle to a vertical line. The amount of inclination varies from 10 degrees to 30 degrees.

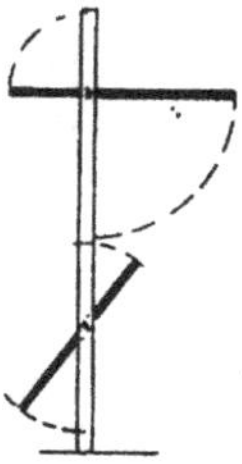

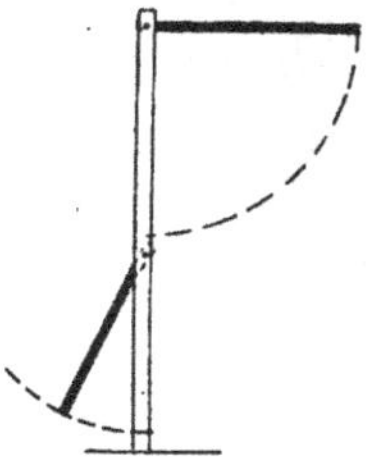

FIG. 352 FIG. 353
Showing the range of action with the joints located in different positions. It is obvious that the locations shown in the right hand figure are the better.

While a sloping shield offers less resistance to the air, this is not the principal reason for the manufacturer sloping it, as it is more a feature of modern style than a mechanical improvement. A car with a sloping shield has the appearance of being able to go fast, just as the sloping lines are in keeping with the knife-like shape that is associated with all devices for fast traveling.

FIG. 354
This double glass V-type shield is being used on some of the custom-made cars.

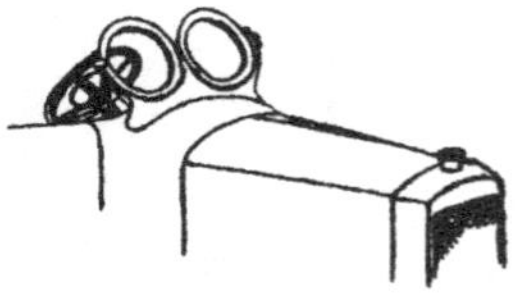

FIG. 355
A "spectacle" type of shield that has been fitted to a few roadster and sports bodies

In Fig. 343 is shown how the air lines impinge on the shield when it is straight, causing a direct interference. The air shoots up, creating a cross draft and an unpleasant back draft in the rear end of the car. In the sloping shield shown in Fig. 343, the air lines flow easily over the windshield and off at the top, blending, to a great extent, with the upper air currents.

There is yet another advantage of the sloping shield, and that is that the length of the top is reduced as shown by the distance marked A in Fig. 344. By cutting off the front end of the top, a better road vision is obtained. This is shown in Fig. 344, by the lines B′B′. The table (page 164) was prepared by M. E. M. Vance and published in "The Automobile" July 26, 1917. It shows the reduction in percentage of the shielded area as a definite sized shield sloped, starting from 10 degrees

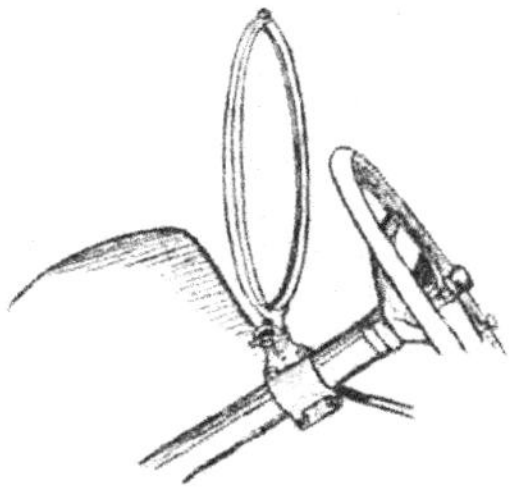

FIG. 356
A modification of the "spectacle" type shield which is seen on a few sports cars. It is attached to the steering column.

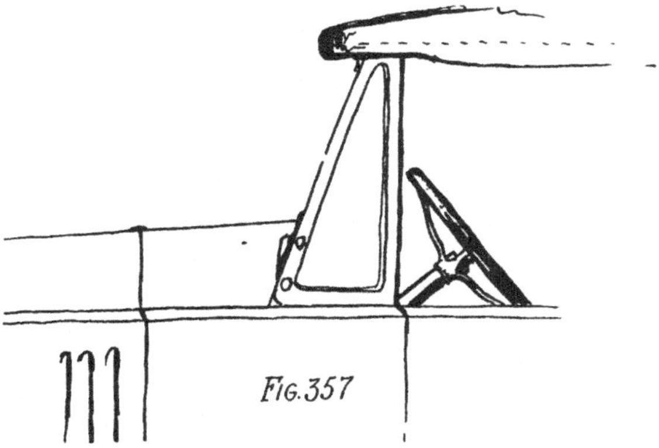

Some cars have small triangular side glass frames on each side of the windshield

and progressing by 2 degrees at a time, until an angle of 30 degrees is obtained. The corresponding reduction in wind resistance is shown for each position. It will be noted that a moderate inclination does not appreciably reduce the projected area of the shield, but it results in quite a reduction of air resistance. The table follows at top of the next page:

Angle of Plane	Reduction of Projected Area (Per Cent.)	Reduction of Air Resistance (Per Cent.)
10	1.5	3.0
12	2.2	4.2
14	3.0	5.8
16	3.9	7.6
18	4.9	9.5
20	6.0	11.6
22	7.3	14.0
24	8.6	16.4
26	10.1	19.2
28	11.7	22.1
30	13.4	25.0

The idea of a sloping shield is not a new one. The Berliet Co., of Lyons, France, fitted a sloping windshield to their cars in 1909, and called it the Berliet wind screen. They developed the shield to have a certain relation to the line of

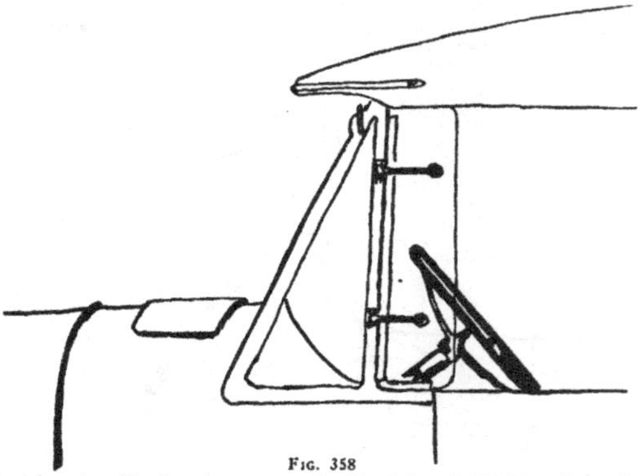

FIG. 358
This car has the triangular side glass frames on each side of the windshield, and an adjustable extension to deflect air currents away from the interior of the body.

sight of the driver, which would cause the air waves to be deflected over the head, relieving eye strain and eliminating the use of goggles. In 1911 they had fitted both rain vision and folding top shields to their cars. In a limousine body they arranged a ventilator in the top of the body, over the driver's head, so that the air would not be deflected on to his head and neck. This is what will happen on an ordinary style of shield. The Berliet Co. description of their shield development is as follows:

"As the speed of the car increases the wind resistance increases as the square of its velocity, which is a very considerable factor in power consumption. For this reason the size of the windshield should be made as small as possible, and its shape should be designed to reduce as far as possible the surface exposed to the wind. These considerations led to the development of a small single piece shield which was set on an angle of about 17 degrees and as closely as practical to the

Fig. 359
A typical design of windshield wing as applied to a rear seat.

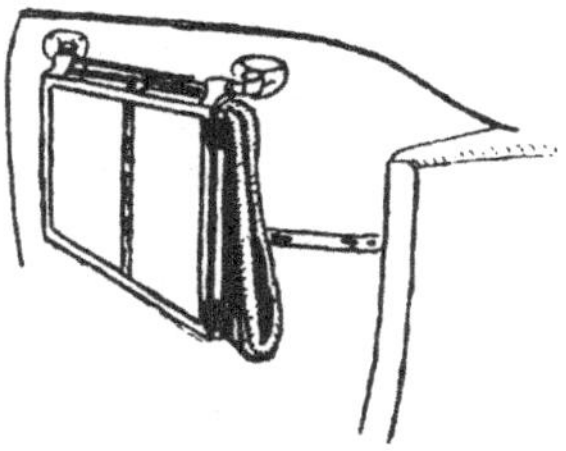

Fig. 360
The rear windshield swings down when not in use.

steering wheel. It was found with this design that the air currents were projected upwards sufficiently to deflect the currents directly in a line with the top of the shield and in so doing did away with the direct draft on the rider's face to a con-

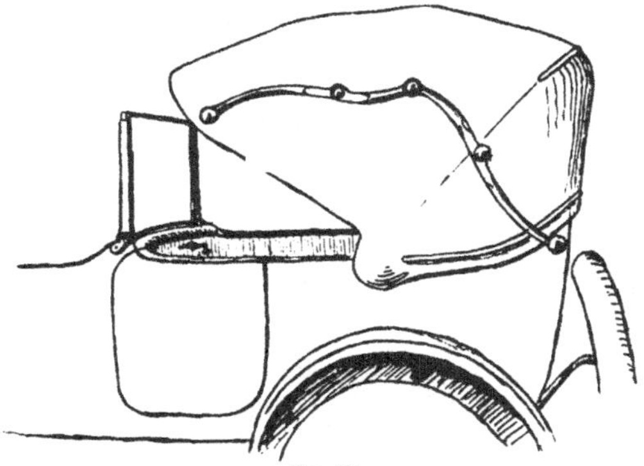

Fig. 361
Single shield designed to protect rear seat passengers in a Victoria top body.

siderable extent. This would only apply in case the top were down. If the top were up, an unpleasant back draft would be created. The drawback of a single shield like this is that it does not afford proper protection in bad weather, so a double shield was evolved which would permit using the upper glass for rain protection in bad weather, and in cold weather it could be closed up tight. In good weather one upper glass is swung out of the way and the lower glass will deflect the air. By either lowering the top or raising the curtain in the back of the top a very pleasant riding condition is obtained. This is illustrated in Figs. 344, 345 and 346.

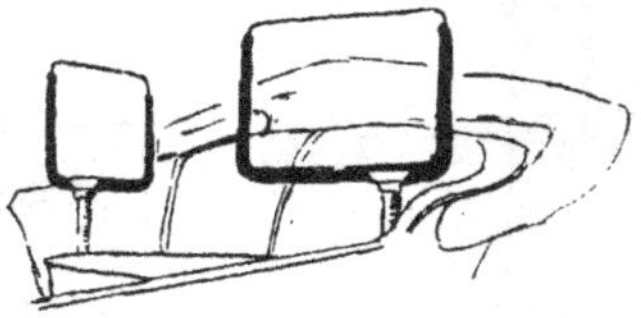

FIG. 362
The shields shown here may properly be called auxiliary devices. They afford only individual protection.

It is seen that the developments along the lines that the Berliet Co. worked has produced the shields that we have for the present day car equipment. In order to complete the study of the windshield the earlier types will be briefly described. The shield shown in Fig. 342 comes in the first group, and shows how the shield was first adapted to the automobile. A filler board was attached to the

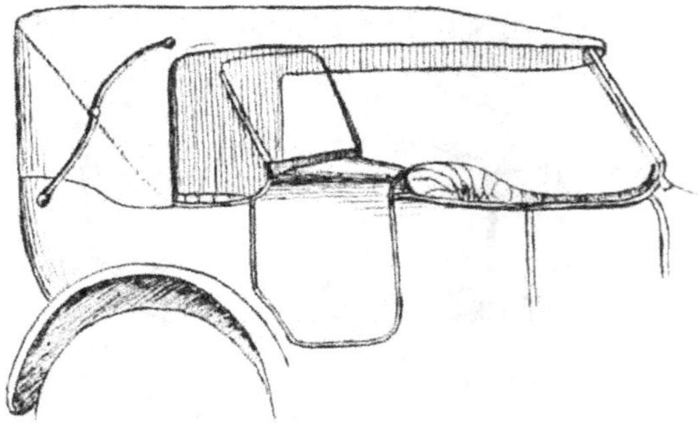

FIG. 363
This shield rests on the end of a cowl-like cover, that fits across the opposite doors of the body. The cover and shield may be folded for ingress and egress. See Fig. 364.

cowl, or an extension was made on the cowl, which had a groove in it that would conform to the curve of the shield frame. This shield was set on to this part and clamped into position by two or more metal clamps. This secured the bottom firmly but would not support the top, so an extension brace was fitted to the joint of the shield. This extension was either carried down to the frame or fastened to the cowl. In Fig. 348 is shown a shield that comes in the same group as the one just described, this shield is used on a cheap car that is made in great quantities. The construction of the shield is essentially the same as that shown in Fig. 342, the only difference being that it is attached by screws and rivets to the corner brackets on the cowl, and that it is supported rigidly enough by these brackets so

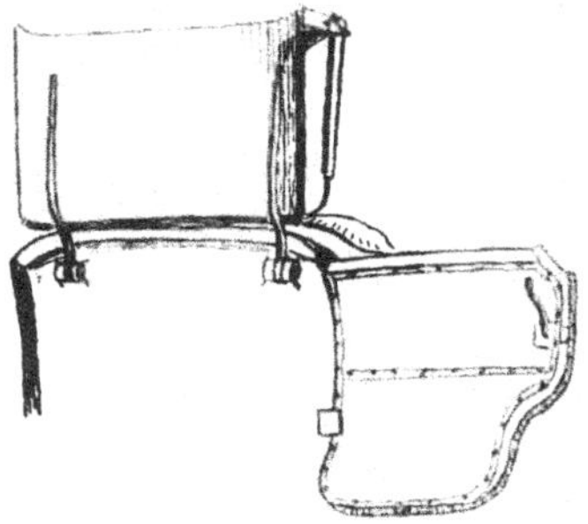

FIG. 364
This shows the cover and shield folded as explained under Fig. 363

that the braces are eliminated. There is another feature, and that is a ventilating condition can be obtained by folding the upper glass back, as shown in the dotted lines. The lower edge of the upper glass projects above the upper edge of the lower glass and so catches the air and deflects it downward into the compartment.

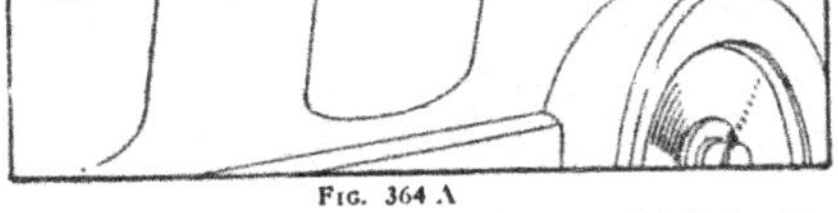

FIG. 364 A
A standard, patented type of rear windshield. (Tonneau Shield Co., New York City.)

The second group is illustrated by the shield shown in Fig. 349. This shield has a friction joint in each of the standards which allows the shield to fold over, as shown by the dotted lines. Each glass frame is also adjusted as they swing on friction joints. To be successful this shield must have a very substantial joint in the post in order to keep it rigid when it is in its upright position. Otherwise the wind resistance and vibration will shake it loose.

The third type is shown in Fig. 350. This is only adaptable to a short shield, as it is very difficult to keep the friction joint tight enough to hold it up for a clear vision.

The shield shown in Fig. 351 is of the type described as "standard." The glass frames have a working range as shown by the dotted lines. A detail of im-

FIG. 365

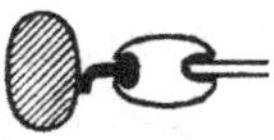

FIG. 366

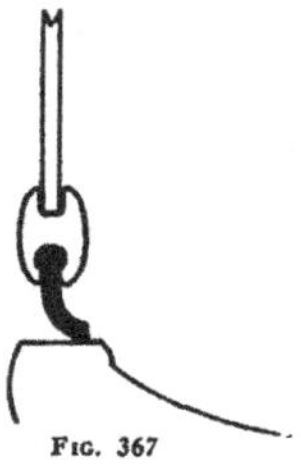

FIG. 367

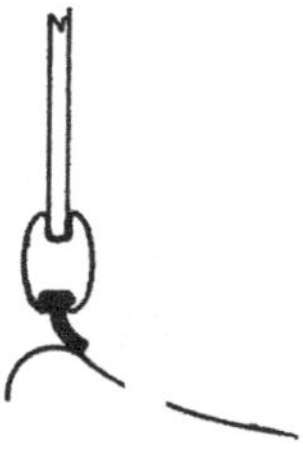

FIG. 368

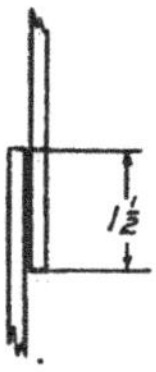

FIG. 369

FIG. 370

FIG. 371

The seven figures on this page illustrate ways of making windshields weatherproof by means of fillers, usually of rubber, between the glass frames and the standards.

portance in the design of the shield is to have the friction joints located properly. The Figs. 352 and 353 show the range of action with the joints located in different positions. It is obvious that the joints located as shown in Fig. 353 are the better, as a more extended range of vision is permitted in the upper glass when it is swung up, and also a better ventilation position is afforded on the lower glass.

In another group are classed designs of shields that are very interesting, as some of these may be the acceptable shields of the future. A V-type shield, shown in Fig. 354 is being used on a number of custom-made cars, the main idea being novelty of appearance and a shield shape that will conform with the car lines. A spectacle type of shield that has been fitted to some special roadsters is shown in Fig. 355 and in Fig. 356 is a modification of this type made in the form of an oval glass that attaches to the steering column.

There are several companies who are making cars with shields that have small triangular side glass frames on each side of the windshield. These are especially adaptable for a sloping shield as it will be seen in Fig. 357. By using

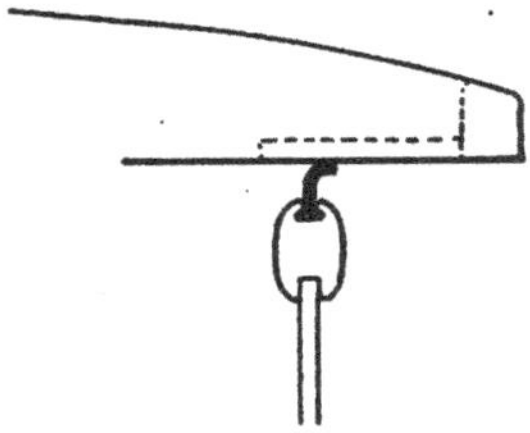

Fig. 372

A weather strip is often fitted in between the top of the windshield and the car top. A close joint should be made the entire length of the rubber strip.

the side shield it gives better vision to the driver especially in bad weather when the side curtains have to be used. It is much easier to get a good fitting side curtain with this shield as the curtain has a straighter line to attach to. The shield construction shown in Fig. 358 has a further extension on it that can be adjusted to deflect the air current so that they cannot enter the car. The common name for these extensions are windshield wings. These wings can be used in connection with any windshield and are very popular. A number of cars are using them for standard equipment. A typical design of windshield wing is shown in Fig. 372B. In Figs. 361 and 362 are shown three types or shields designed to protect the rear seat passengers of a touring car. In Fig. 359 is shown a shield that attaches to the back of the front seat. This swings down out of the way when not in use, as shown in Fig. 360. When it is in operation it can be set at any desired position by sliding it back and forth on two rails. Side glasses are provided to afford protection from side winds. In Fig. 361 is illustrated a shield that can be lowered into the back of the front seat when it is not wanted. It is operated by a handle which raises or lowers the shield as desired. When the shield is down there are two flaps that fold down and cover up the slot in the seat. The shields shown in Fig. 362 may be called auxiliary devices, as they afford only individual protection.

A tonneau windshield is an efficient device for protecting rear passengers, and it is surprising that it has not been used more extensively. A protecting shield for the rear passengers will be a part of the standard equipment before very long. Another tonneau shield of interest is shown in Fig. 363. This consists of a shield on the end of a cowl-like cover, that fits over the doors of the body. This is hinged at the rear of the front seat, and is folded back when the rear passengers wish to get in and out. A car fitted in the same manner was exhibited by the Alin & Liautard Co. in 1914, and has been duplicated by special body builders on some recent American models. In Fig. 363 is shown the shield in position on the car, and in Fig. 364 is shown folded up to allow the passengers to enter and leave the car. It will be noted that the glass is adjustable so it can be set in any desired position.

In 364 A is pictured the "J. H." tonneau windshield, which has become exceedingly popular.

The automobile has ceased to be a fair weather vehicle, and consequently the demand is for complete protection from bad weather conditions and the cold. In order to make the windshields completely weather-proof, rubber fillers are being

FIG. 372 A
A type of "three-way" shield that is extensively used on large production as well as custom made cars. (Joseph N. Smith & Co., Detroit.)

fitted between the glass frames and the standards. The shapes and sizes of these rubber fillers depend upon the shape of the shield. Sometimes they are attached to the glass frames and sometimes to the standard. (See Figs. 365 and 366.)

With the advent of the swinging lower glass a weather strip was made necessary; two cross sections of typical weather strips for the lower glasses are shown in Figs. 367 and 368. Several methods have been devised to make a weather-tight joint between the upper and lower glass. The first system was to have the upper glass overlap the lower one, as shown in Fig. 369. This works fairly well, but does not keep the cold air or rain from blowing up between the glasses. In Fig. 370 is shown a section of the upper and lower glasses with a rubber filler inserted between them. This method is used very extensively, as it makes a good close joint between the glasses. There is one trouble, and that is to get the shields to have a uniform space between them as any distortion of the body reduces or increases this space, and when this gap is reduced it is liable to cause a fracture in the glasses.

The system shown in Fig. 371 overcomes these objections, and the only factor that is to be watched is to see that the glasses are set square in the standards so that a close joint is made along the entire length of the rubber. On the top of the shield a rubber weather strip is sometimes fitted, and this will work very well when a level surface is provided on top for this strip to fit against. This condition is shown in the Fig. 372. The weather strip is attached to the frame in the same

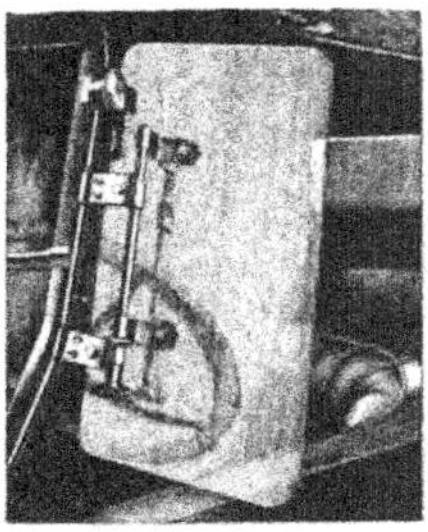

FIG. 372 B.
A typical design of windshield wing.
(Automotive Accessories, Inc., Kokomo, Ind.)

manner that it is fitted to the bottom of the lower glass frame. When it is raining or snowing the glasses of the shields are quickly fogged, so that it is difficult to see through them. In order to get any vision the shield is opened by raising the upper glass upward and outward to the so-called rain vision position, making it possible to see between the two glasses. This method is satisfactory providing it is not raining very hard, and also, in case of snow, the cold air coming in is not objectionable.

In order to avoid the inconveniences that an open rain vision shield causes, a wiping attachment can be used. This device is fitted so that it can be operated from the inside of the car and cleans off the rain or snow from the outside glass by a backward and forward movement of the hand. Automatic windshield wipers are also on the market, as shown on pages 137 and 138.

TOP DESIGN

CHAPTER XVII

A collapsible top is used on the greater number of the automobiles of today, and while there is an ever-increasing demand for the closed type car, the touring car, with a folding top, still retains its popularity. A folding top on a car gives a freedom to the passengers that is second only to being out walking, and when country travel is indulged in in fine weather, the pleasures are greatly increased if the top can be folded away, giving an unobstructed view in all directions.

The first tops used on the automobile were adaptations from the horse-vehicle tops, and as the automobile tops were long, curtains had to be used on the sides

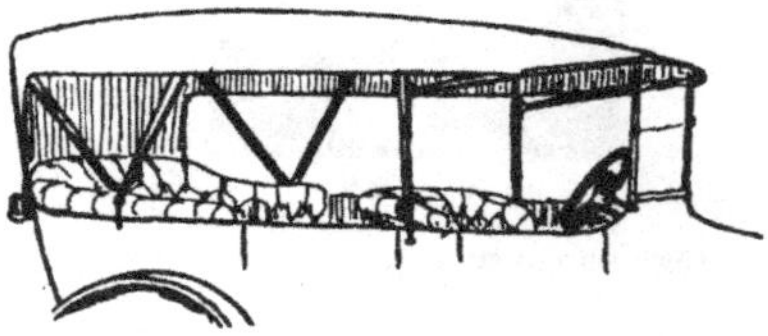

Fig. 373
Typical design of top common before the general adoption of the "one-man" type.

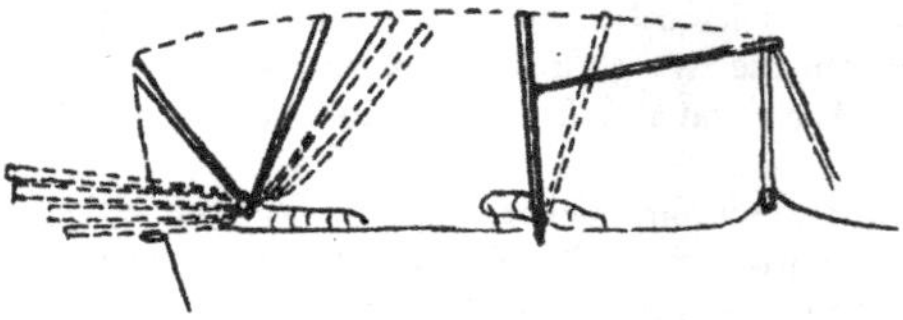

Fig. 374
This top consists of two front and two rear socket assemblies, right and left. The sockets are made of steel and so shaped as to accommodate the wood bows driven into them.

of the top to afford proper protection. The developments of the earlier tops were not of much interest and a detailed description will not be taken up.

The top shown in Fig. 373 is a typical design of top that was commonly used previous to the adoption of the "one-man" type. The top illustrated in Fig. 374 consists of two front socket assemblies, right and left, and two rear assemblies, right and left. These sockets are made of steel and so shaped as to accommodate the wood bows that are driven into them.

In this top four bows are used. The front and rear sockets are set on to the top irons provided on the body. A water-proof cover is laid over the top of the

FIG. 375

Four-bow type Du Quesne socket

bows and then stretched tight by attaching the rear bow to the back of the body by two straps, and at the front two straps run from the front bow to convenient places on the chassis or body. These straps are pulled up tight so that the covering will be well stretched and so make a smooth appearing top.

To close the top down, the front sockets have to be taken off the body iron and attached to the front member of the rear socket. This is accomplished by

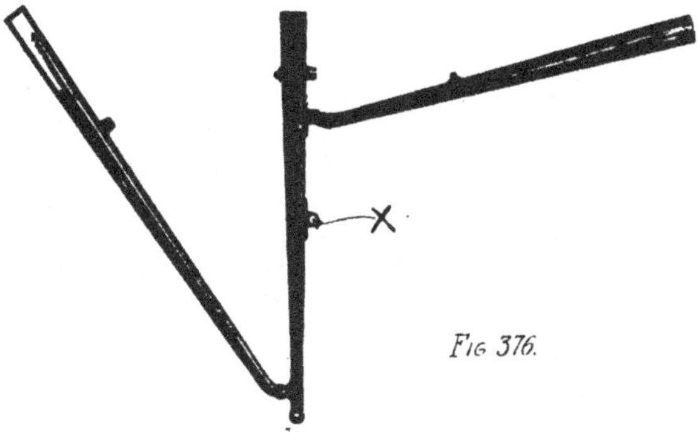

FIG. 376.

Roadster type Du Quesne socket.

having a bracket attached to the sockets that has a hole in it the same size as the front body iron has. The pin on the end of the front socket will then fit into this at the rear, the same cotter pin being used to hold it in position as is used when the top is in its open position. This is shown very plainly in Fig. 374.

These sockets are known as the Du Quesne. Photographs of the sockets are shown in Figs. 375 and 376. Fig. 375 shows a four bow type and Fig. 376, a roadster type. In this type the front bow comes off and attaches at point marked X, when the top is folded down. A strap is riveted to the socket as plainly shown; this carries the cotter pin which keeps the socket in position.

To operate this top successfully it was necessary to have one person on each side of the car. Each person would take hold of a front socket so that it would be raised or lowered evenly. On a large top it was difficult for even two men to manage. If a sudden storm came up the chances were that the passengers would be thoroughly wet before they could get the top up.

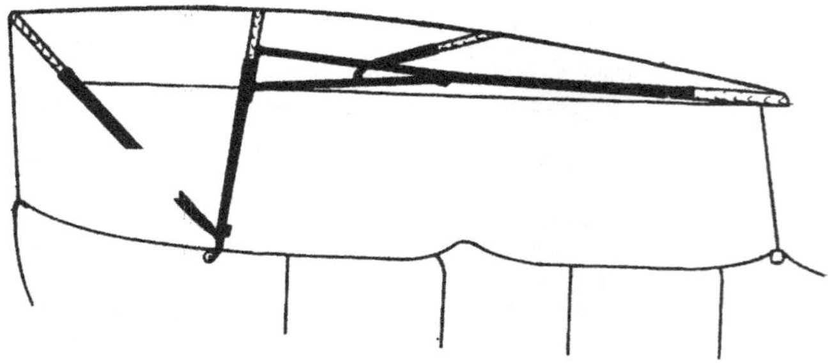

Fig. 377
Typical layout for bow sockets.

In order to overcome these objectionable features socket constructions were worked out where the front end was supported by suitable members from the rear bows. The front members were so designed that they would balance one another when the operation of raising or lowering the top was performed. The theory was that one person should be able to raise or lower the top while in the car.

In the first designs of this "one-man" type top the construction was prone to run to a good many levers and cables, which gave trouble by catching in the cloth and getting tangled in operation. When the construction was developed to work easily it was found difficult to get the desired outline on the top, as the lever and bars had to maintain certain definite relations to one another.

To meet the conditions demanded, further movements had to be developed and the parts simplified. This has resulted in the type of sockets that are now most generally used. In Figs. 375, 376 and 377 are shown three typical layouts for sockets. They are all used very extensively. It will be seen that they all have a truss construction while the covering forms the tension cords. The rear end is anchored by straps to the back of the body while the front end is secured fast to the windshield.

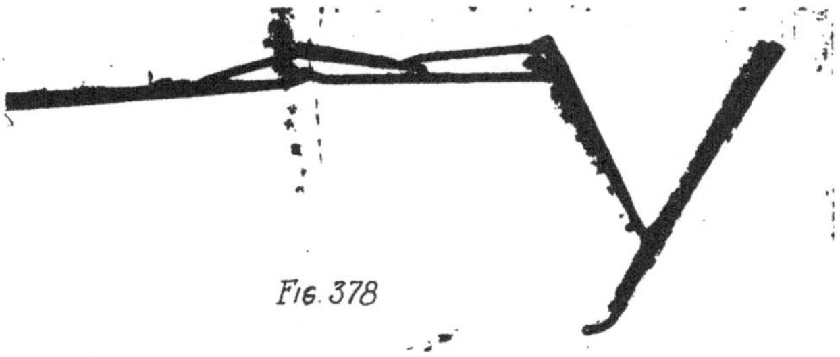

FIG. 378

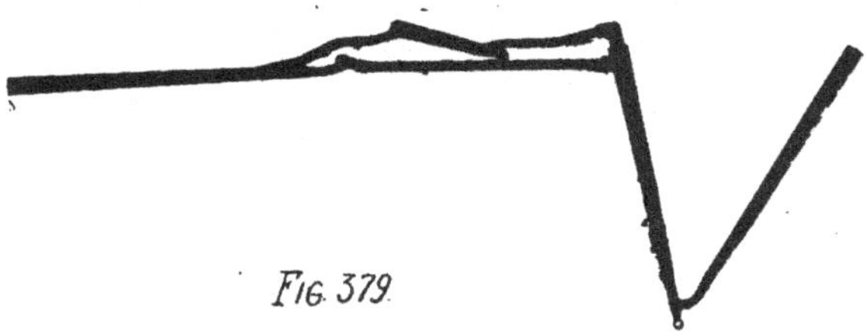

FIG. 379

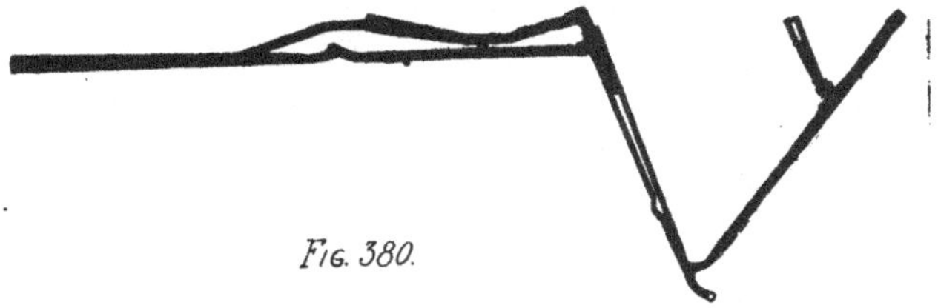

FIG. 380

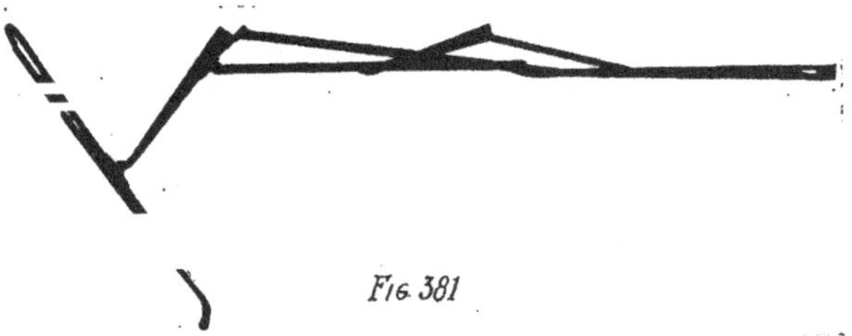

FIG. 381

Typical designs of bow sockets whose points of interest are referred to on page 176. (Manufactured by the Brewer-Titchener Corp., Cortland, N. Y.)

TYPES OF MODERN BOW SOCKET CONSTRUCTIONS

In Figs. 378, 379 and 380 are shown types of modern socket constructions. Figs. 378 and 379 are four bow sockets and Fig. 380 is a five bow socket. It will be noted that the third bow in socket Fig. 378 is attached to the fourth while in Figs. 379 and 380 the fourth socket is fastened to the third. When the third bow is attached to the fourth more space is allowed in the side of the top than there is when the fourth is attached to the third.

The bow sockets shown in Figs. 381 and 382 is a slightly different construction than those shown in Figs. 378, 379 and 380. The braces are longer and the lower strut is made of two pieces which allows the No. 2 socket to set between it. The manufacturing name for this type is the Pioneer. Fig. 382 shows the position the sockets take as they are folded up. Roadster three bow sockets are shown in Figs. 383, 384 and 385. The Pioneer type is shown open and closed in Fig. 383 and 384 respectively. The main difference between the Pioneer type and the one shown in Fig. 385 is the location or point of attachment of the No. 2 bow. In Fig. 385 the No. 2 bow is fastened near the top of the socket while Fig. 383 is attached half way down. This makes a difference in design of the side quarters as indicated by the dotted line.

The top illustrated in Figs. 386 and 387 has a socket which has recently been developed. It has a spring inserted inside one of the tubes. This spring applies pressure through the levers and members so that the operation of raising and lowering the top are assisted to a great extent. This spring makes the top almost automatic, as it raises the top and also prevents it dropping too quickly when it is let down. This principle applied to a long and heavy top makes its operation easy to a remarkable degree.

When an automobile top is designed there are a number of conditions that have to be considered, and these are outlined in the following:

The first factor is to decide what head clearance is desired. This is determined by the distance from the cushion to the point in the top that will come directly above the passenger's head. This is shown in Fig. 388, where a dimension B for the front seat calls for 38-38½, while on the rear seat the letter (A) is from 38 to 40. It can be estimated that the cushion will compress at least 1 inch when the passengers are on it, but if the cushion is an extra soft one and allows the passengers to sink in deeper, the necessary allowance should be made.

The windshield height is the next thing to be decided upon, as the height of the windshield determines the amount of road vision the driver and passengers will have. The trend of design is to get the top as low as possible and also have the windshield short, but these should not be brought down too low, otherwise the driver's and passengers' vision will be obstructed. With these points established the outline of the top can be laid in. The back is set in the desired relation to the body line and at the front the line will terminate to suit the windshield.

The shape desired at the front and rear are taken up next, and as the back presents the larger surface to the eye it will be worked out first. The shape of the

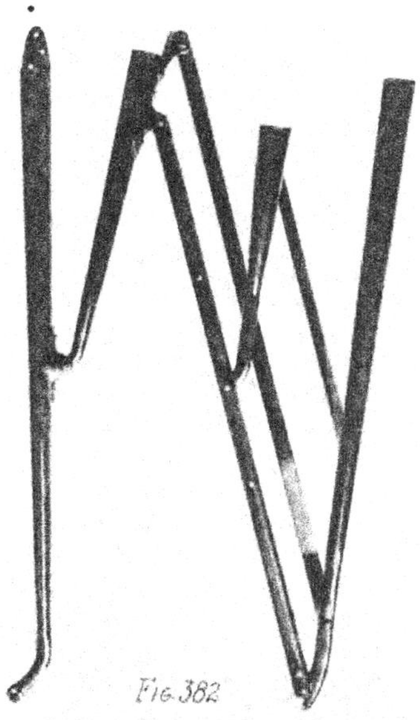

Fig. 382

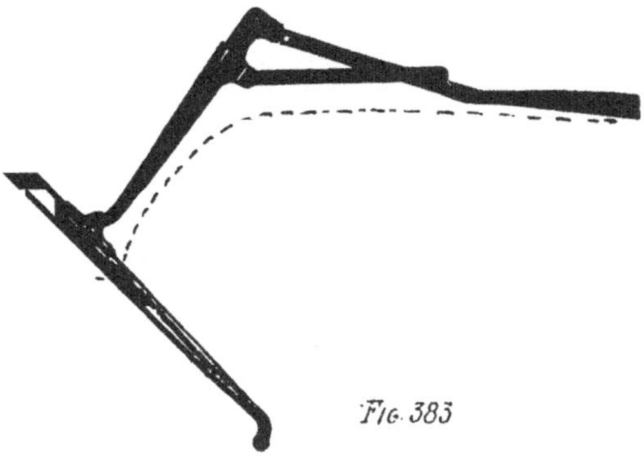

Fig. 383

Typical designs of bow sockets whose points of interest are referred to on Page 176.
(Manufactured by the Brewer-Titchener Corp., Cortland, N. Y.)

back bow and the height at which it sets determines the appearance of the back, so the bow outline is laid in as desired. The other bows are so shaped that they will carry a uniform degree of outline throughout the complete length of the top. The importance of the shape of the top when it is up has been emphasized and the next consideration is to insure that it has the correct shape when it is folded or let down. The bows must be wide enough to clear the body and they should not project too far at the rear of the body. The line that the top is set in when it is folded should conform to the general lines of the body design. It is common practice to carry the spare tire at the rear, and when this is the case care has to be exercised to see that there are no bows setting over the tire so it would interfere with getting the tire on and off the tire carrier.

The best way to arrive at satisfactory results is to make a preliminary layout and then submit this to the company which is to make the sockets, and as there are peculiarities in each socket manufacturer's constructions, it is advisable to have them go over the design and see if their constructions can be adapted to the requirements.

The bows used are made from ash or oak, which is machined and bent to the shapes required. While various sized sections are used, the 1½ by 1 is the most popular. The ends that are to be driven into the sockets are shaved off so that they will conform to the taper end of the socket, as shown in Fig. 389. In specifying the shape of the bow to the bowmaker, the width, height, section, radius and crown are required. The dimensions A, B, C, D and E required are shown in Fig. 390, and when they are given in the same manner as shown the bow-maker will have no trouble in understanding the requirements. .

The different sections shown in Fig. 391, A, B, C and D, give a good idea of the shapes that are required for the bows used at different parts of the top. These sections show the corners of the bows removed on the edge or edges that come in contact with the covering. If these corners were left sharp the edges in time would cut through the top material. In Fig. 392 is shown how these corners set in the top.

The use of a tubular socket is standard American practice. The European practice is to use a slat iron or bracket construction and the bows are attached by wood screws. By making the top construction this way the parts are made slightly smaller and in some cases neater, but the cost and time required to make them offset any advantages they may possess.

A typical wood bow and bracket socket construction is illustrated in Fig. 391-A. This car has the top bows with a natural finish and the brackets or ends bronze, nickle-plated. The photograph does not show the sockets to advantage, but they look much better than the enamel steel sockets, when the sockets are exposed. The general trend is to have the top material cover the outside of the sockets.

On the front bow a suitable fastener has to be attached to secure the top to the windshield, and as there are several different methods used to do this it will be interesting to review them. The most common type of fastener is one similar to that shown in Fig. 393, which consists of a malleable iron bracket with a screw that has a convenient shaped head, preferably with winged extensions. A sectional view in Fig. 394 shows how this set screw engages with the pin on the windshield. Another type along similar lines to that just described is shown in Fig. 395. On this fastener the winged head screw clamps a ring on the ball end of the shield.

In Fig. 396 is shown a toggle lever arrangement for clamping the top to the shield. The fastener shown in Fig. 397 is a semi-automatic, as the slide is operated

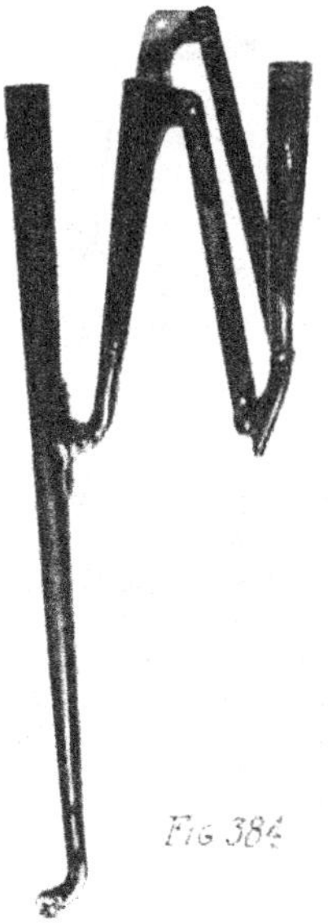

Fig 384

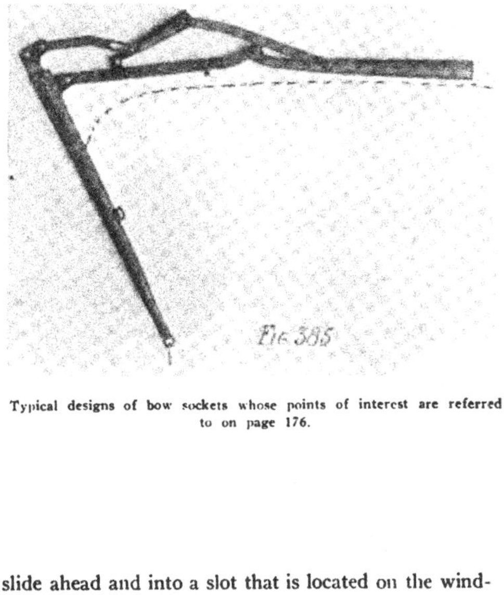

Fig 385

Typical designs of bow sockets whose points of interest are referred to on page 176.

by a spring that pushes the slide ahead and into a slot that is located on the windshield bracket as shown in the sectional view. To disengage the top this slide is pushed inward, which disengages the slide from the windshield post, leaving the top free to rise. This is a very simple arrangement and when it is properly fitted works in a satisfactory manner.

The back bow has to be braced in position by back brace straps. There are several ways of arranging these straps; some of these methods will be described.

The first method is to use a double strap that is looped into a suitable fastener. The loose ends of the strap are nailed to the bow as shown in Fig. 398. The second method is to attach the straps to the top of the bow and let them come down over the inside of the bow as shown in Fig. 399. The third method is to use a single strap with a buckle to allow for adjustment as shown in Fig. 400. A fourth style, shown in Fig. 401, is to use a double end of strap that is Y-shaped where it attaches to the bows. This can be fastened either to the inside or outside of the bow. Style number five, shown in Fig. 402, is to attach a loop to the bow

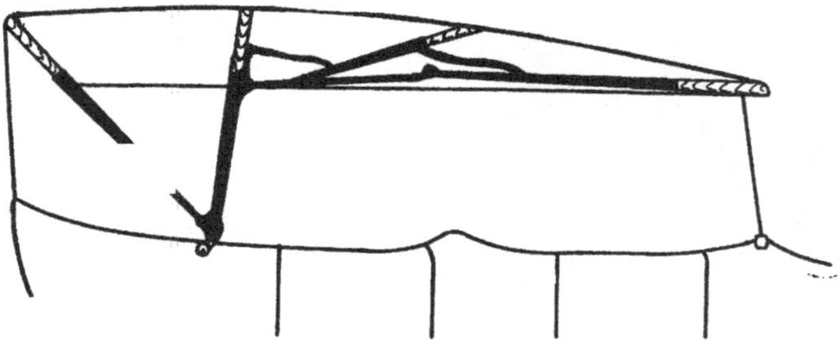

FIG. 386
This top has a socket which has recently been developed. It has a spring inserted inside one of the tubes. The spring is shown in the figure below.

and put either a leather strap or cotton strap through it and the fastener on the body, the ends of the strap to be fastened together with a suitable buckle.

The back and side curtains will next be described. The back curtain attaches to the top of the rear bow and to the back end of the body. It is generally made of one piece, the upper end being nailed in position and the joint it makes with

FIG. 387
The spring referred to under Fig. 386, applies pressure through the levers and members so that the operations of raising or lowering the top are assisted to a great extent.

the top cover is covered by a suitable molding, while the lower end is attached either by nailing or by suitable detachable fasteners as shown in Fig. 403. There are some designers who like to have the back curtain in three pieces, as shown in Fig. 404, in order to permit the center section to be raised in hot weather. The reason for raising this portion is to permit the air to pass right through the car and not have any back drafts.

Another style of back curtain which is largely used has extensions on each side that run to the back bow socket. These are called gypsy curtains. They protect the rear passengers to a great extent from the dust. In the design shown in Fig. 405 these gypsy sides are extensions of the back curtain. There is another method of making these, and that is to put small, separate triangular curtains in

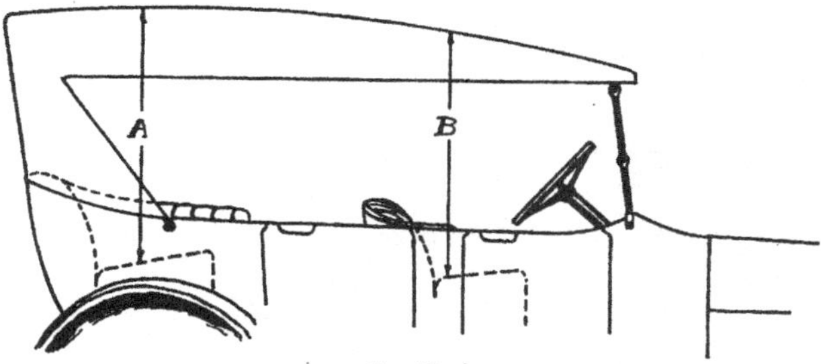

FIG. 388
Method of determining head clearance in designing an automobile top.

and fasten them to the back curtain and bow socket as shown in Fig. 406. While this is not as good a job as making them integral with the back curtain, they answer the purpose very well in providing protection.

FIG. 389
Bow ends that are to be driven into the sockets are shaved off so they will conform to the taper end of the socket.

A suitable light has to be provided in the back curtain. This can be made by cutting a hole in the fabric and sewing a piece of celluloid over it. In putting in a celluloid light several small lights are used more practical than one large one

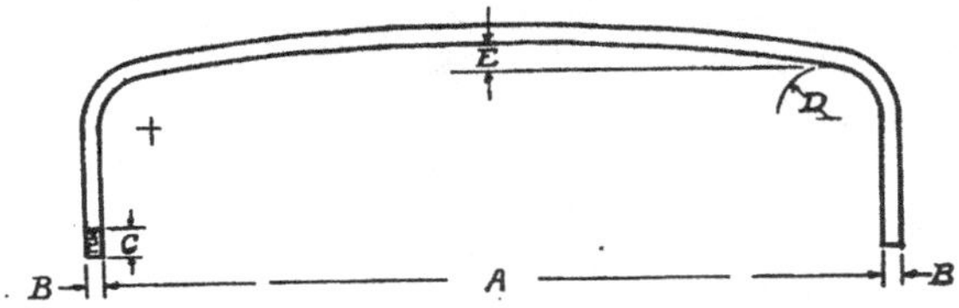

FIG. 390
The dimensions A, B, C, D, E, when given in the same manner as here shown will give the bow maker a thorough understanding of the requirements.

because the celluloid hardens and cracks in service and the larger the light the more trouble it will give. Back curtain lights can be made in all sorts of shapes and in Figs 407 to 412 are shown some of the designs that have been used. Back curtain

A typical wood bow and bracket socket construction is illustrated in this photograph. The car has the top bows with a natural finish and the brackets or ends bronze, nickel-plated. Of course, the photograph cannot show the sockets to advantage, but they look much better than enameled steel sockets.

lights made of glass are becoming very popular as they are more ornamental and give better vision than the celluloid ones do. The glass light has one drawback, and that is its weight, consequently the sizes of the lights are smaller than when celluloid is used.

The light with the rounded end like that shown in Fig. 413, and the rectangular light shown in Fig. 415, are good, conservative shapes which can be worked in

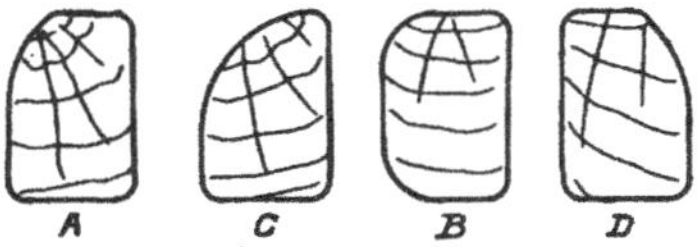

FIG. 391
The different sections A. B. C. D. give a good idea of the shapes required for the bows used at different parts of the top.

practically any design and would be good shapes for the curtain builder to carry in stock. They should be about 6 x 20 or 8 x 24 for the general run of work, as they give a good rear vision in these sizes.

In making the following sketches, Figs. No. 416 to No. 427, of back light shapes, no attempt is made to vary the shape of the back to harmonize with the different combinations. These sketches are merely a collection of light designs arranged for the benefit of designers or students, and while it may seem that the draftsman tried to lay out as many designs as could be conceived, they are all

FIG. 392
This drawing shows how sharp bow corners set in the top.

taken from actual tops that have been built, and in no cases are they merely imaginative designs. The aim of each top builder has been to create a back light that is different from the existing lines.

Those lights shown in sketches which have sharp corners like Figs. 418, 420, 421, 422, 424 and 425, are all hard to make, owing to the difficulty of fitting the glasses. Then again the sharp corners are apt to localize the vibration and cause cracks. From the standpoint of vision, of course, they are of no benefit and in some cases where bevel glasses are used will cause unpleasant light reflection. If a quantity of tops are to be made it will be advisable to choose a simple design, similar to those shown in Figs. 413 and 415, as it will be a great time and labor

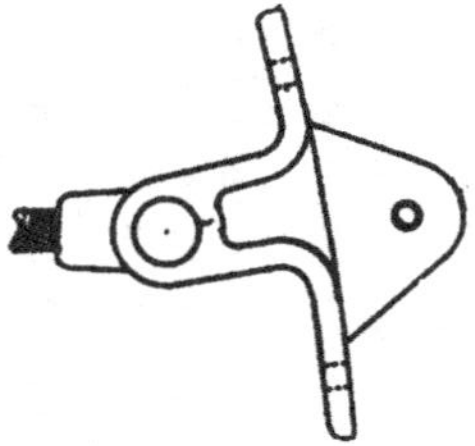

Fig. 393
A common type of front bow fastener.

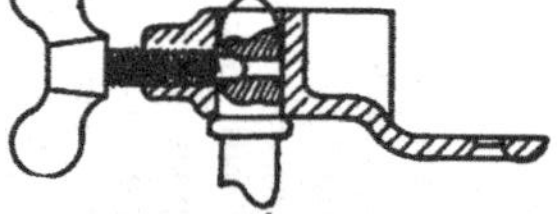

Fig. 394
Sectional view of above figure.

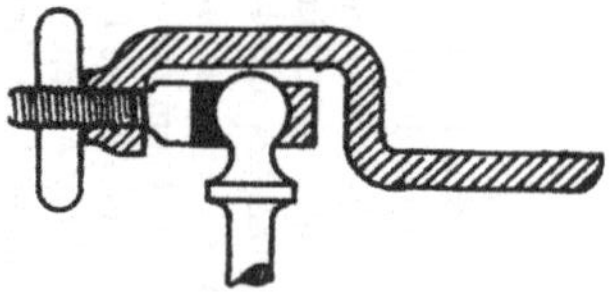

Fig. 395
Another type of front bow fastener.

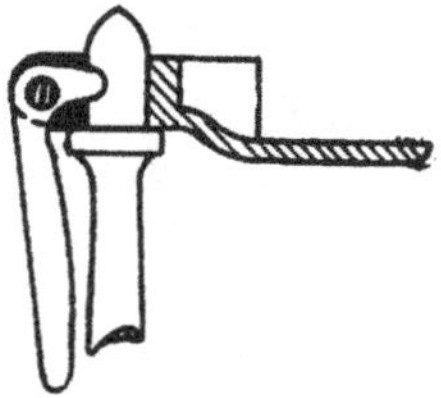

Fig. 396
Toggle lever arrangement for clamping the top to the shield.

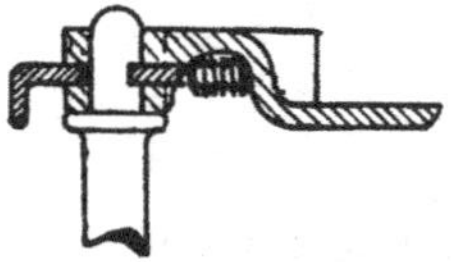

Fig. 397
Semi-automatic front bow fastener.

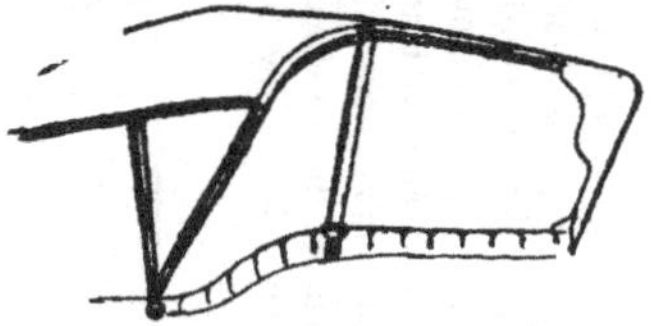

Fig. 398
A double strap is looped into a suitable fastener. The loose ends of the strap are nailed to the bow.

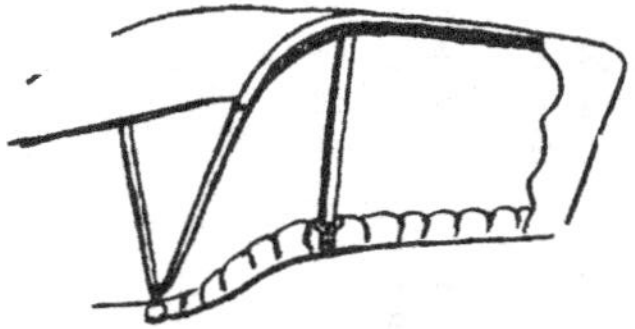

Fig. 399
The second method is to attach the straps to the top of the bow and let them come down over the inside of the bow.

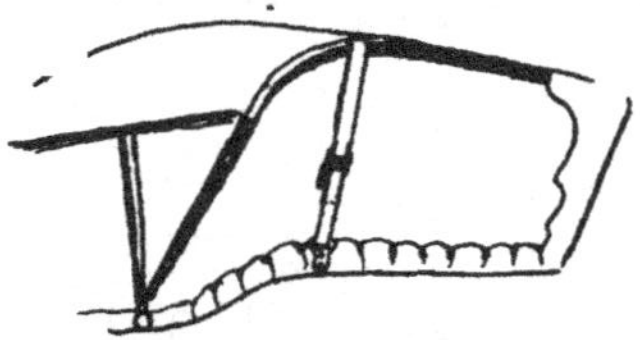

Fig. 400
The third method uses a single strap with a buckle for adjustment.

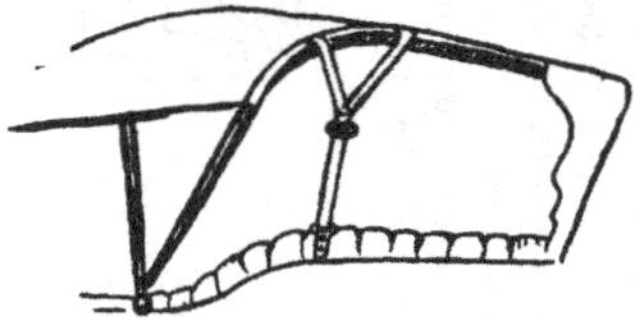

Fig. 401
The fourth method uses a double end of strap that is Y shaped where it attaches to the bows.

saver, both in cutting the glass and making the frames. In addition to this they make a much more simple job when it comes to fitting the fabric on around the frames. One of the main troubles with these back lights is to keep the water out, and it is obvious that it is a great deal easier to make a tight fitting glass in a simple frame than in one of the more complicated shapes.

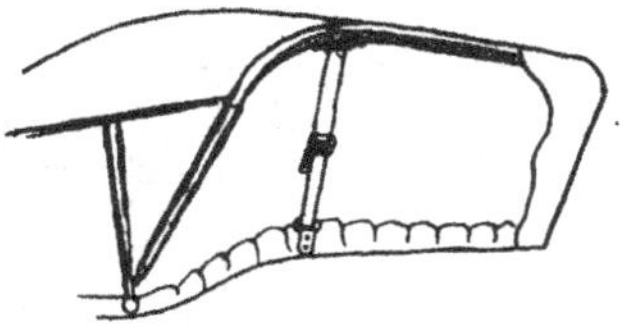

FIG. 402
The fifth method is to attach a loop to the bow and put a leather or cotton strap through it and the fastener on the body.

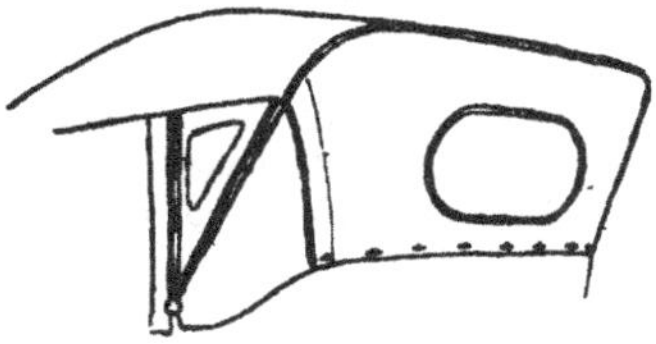

FIG. 403
One piece back curtain.

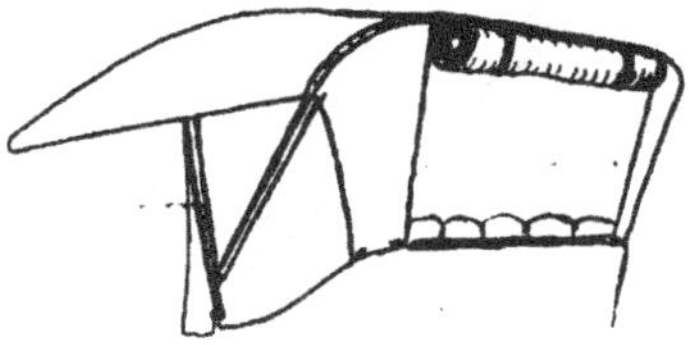

FIG. 404
Three piece back curtain, permitting center section being rolled up in warm weather.

In studying the various light shapes it is apparent that the long, narrow form of light is the predominating type. The only reason for having a light in the rear is to allow the driver and passengers to look out, and to best accomplish this result would be to have a large back light as in the closed cars. This would be quite heavy and expensive, so the next best thing is to have a long, narrow glass, set at a convenient height. A large, round glass like that shown in Fig. 428 would not allow the side passengers in the rear seat or the driver to see out of it.

Some designers like to put in double lights like those shown in Figs. 416 and 417. This is better than one small center light, as it permits the driver and passengers a wider range of vision. The oval type of light shown in Fig. 417 is the most popular and gives a better vision than the round one.

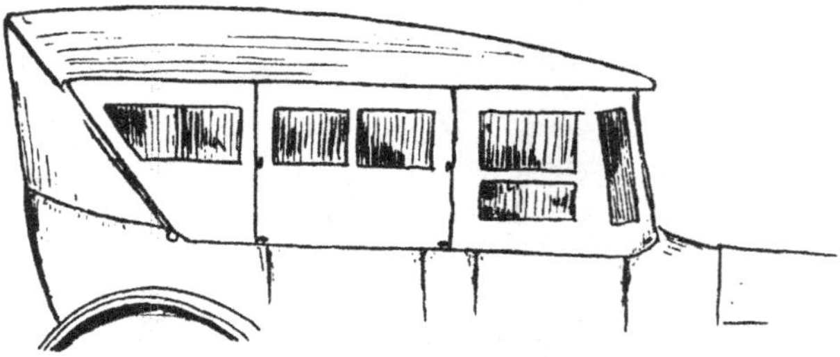

Fig. 405
Back curtain with extensions running to the back bow sockets. These are called "gypsy" curtains.

Having examined the general design of the lights it will be well to take up the various constructions that are used to make them. First of all, the commercial construction will be discussed. Under commercial construction comes the standard light frames that are made and kept in stock by the jobbers. There are

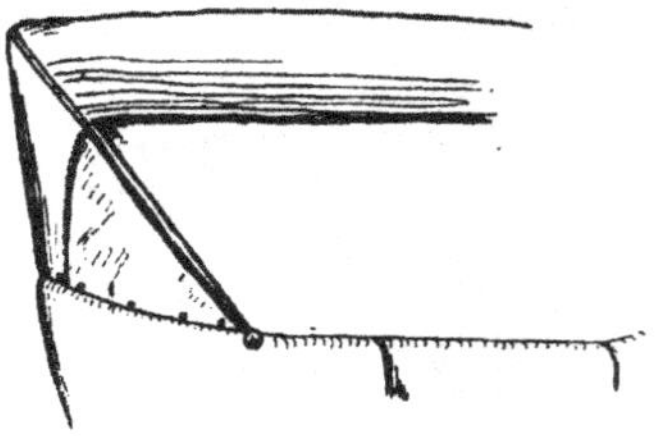

Fig. 406
Sometimes "gypsy" curtains are put in as separate, triangular curtains

a number of companies which make back and side light frames, some of the best advertised makes being the Johnston, Brewer-Titchener, Soss, Perfection and McAvoy.

While the manufacturers have a wide assortment of designs, they necessarily limit their lines because of the great expense in getting the dies and tools made for each design.

The Johnston light consists of three sections, the glass, metal or outer frame, and the inner frame of wood. The inner lining and outer covering are nailed to the wood frame and then the metal frame is set on the wood and fastened to it by machine screws. This clamps the outer covering of the top securely to the wood frame and also provides a retaining ledge for the glass. This is shown very clearly in Fig. 429. The glass is shaped into place and a retaining strip or bind-

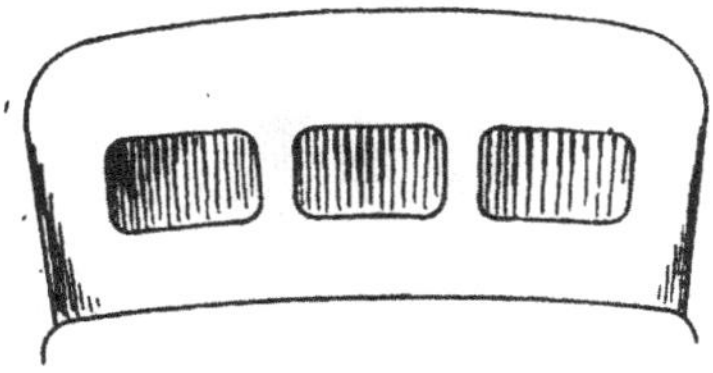

FIG. 407

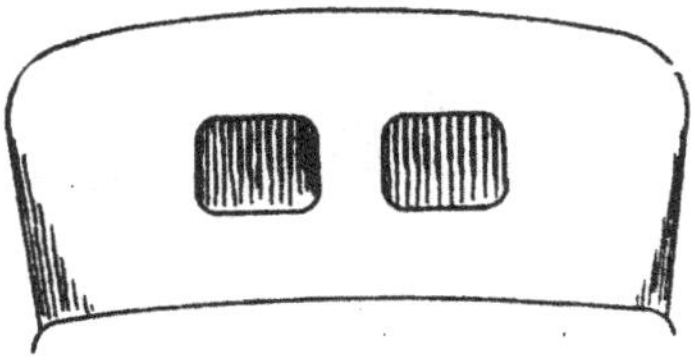

FIG. 408

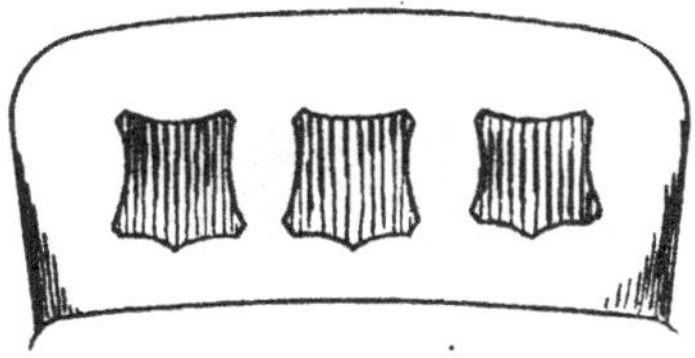

FIG. 409

These three drawings and the ones on page 188, show different shapes and arrangements of back curtain lights.

ing is nailed in place. This strip can be made of rubber composition, covered with cloth, cane covered with cloth, or strips of linoleum covered with cloth. A light brad is used to attach this frame.

The Brewer-Titchener style light is composed of three units, a metal outer and inner frame and the glass. Studs are riveted in the outer frame and these are projected out far enough to go through the inner frame, which is secured by means of nuts turned on the studs. In applying these lights a large hole is cut in the fabric to match the glass opening, and also a number of small cuts are made to match the studs in the outer frame.

The fabric is now laid in the outer frame, then the glass is put in place and the inner frame put on, and when these nuts are tightened down they clamp the

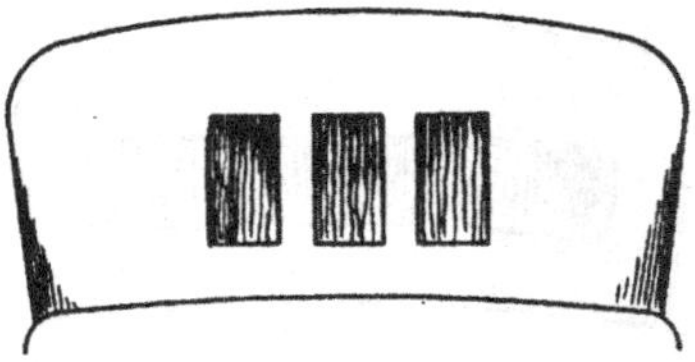

FIG. 410

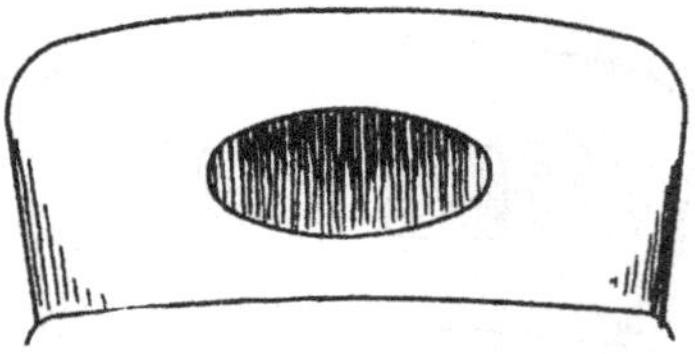

FIG. 411

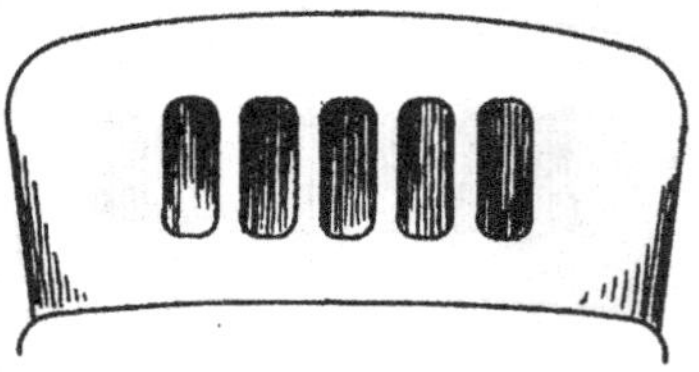

FIG 412
These three drawings and the ones on page 187, show different shapes and arrangements of back curtain lights.

frame, glass and fabric securely. A similar procedure is necessary on the McAvoy lights, the only difference being that these are aluminum alloy castings and are held together by screws that go through the rear frame and turn into the outer frame. This is shown in Fig. 430. Projections are made on the outer frame that match corresponding depressions in the inner frame. These are used to pinch the fabrics when the frame is tightened up.

On the lights described there is a bright finished rim shown on the outer frame. If a smooth finish is desired, a wood frame like that shown in Fig. 431 can be used. This frame is shaped to the desired outline and is thick enough to hold the glass and still leave a ledge wide enough for the retaining binding. To

Fig. 413

Fig. 415

The light with the rounded end like that shown in Fig. 413, and the rectangular light shown in Fig. 415, are good, conservative shapes for the curtain maker to carry in stock. They can be worked into practically any design.

put this light in the back, the fabric is cut out and turned inside the frame, and then the lining is also brought up and tacked into it. When this is done the glass is slipped in and the retaining binding put in place.

The side curtains are made from a material to match the top covering and back curtains, and so arranged that they can be opened conveniently above the

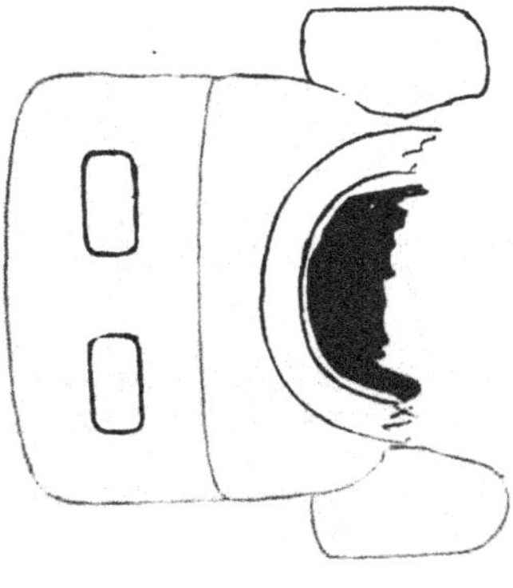

Fig. 416

Fig. 417

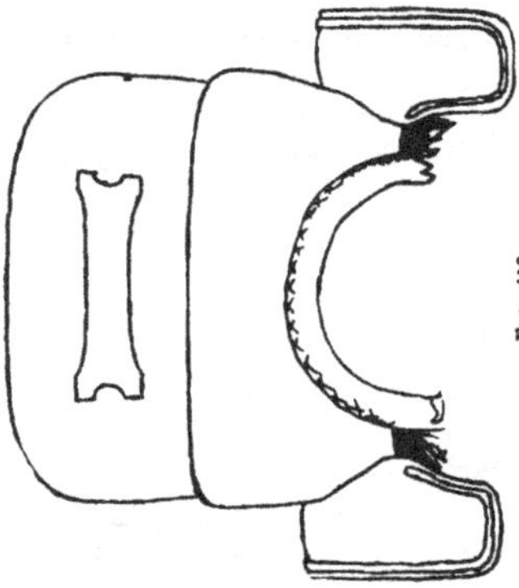

Fig. 418

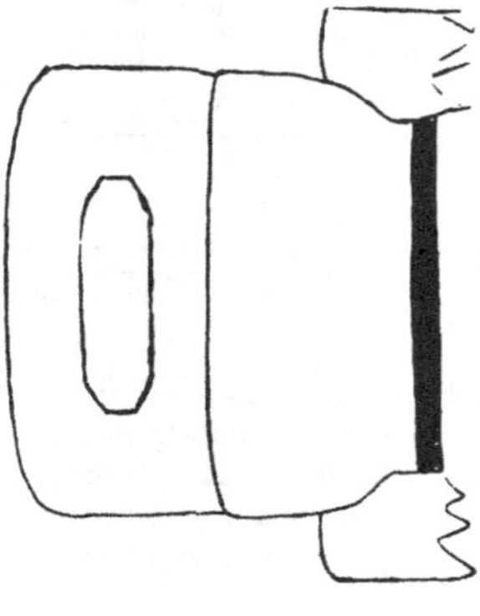

Fig. 419

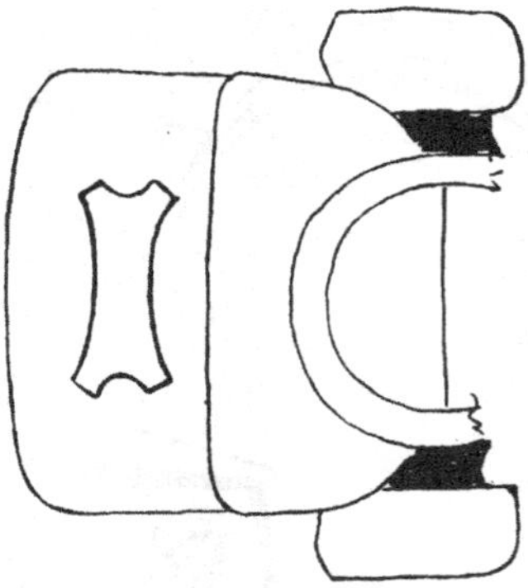

Fig. 420

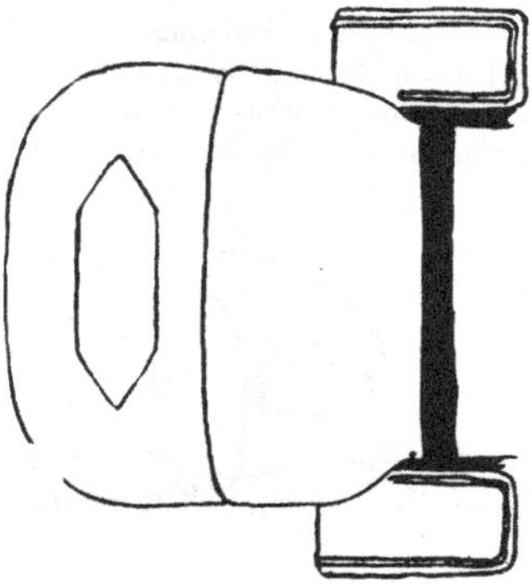

Fig. 421

A collection of rear light designs arranged for the benefit of designers or students. They are all taken from designs that have been actually built. (Figs. 416 to 427 inclusive.)

Fig. 422

Fig. 423

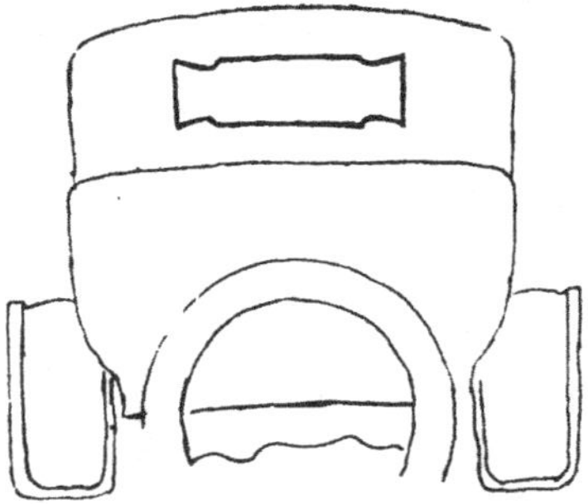

Fig. 424

Fig. 425

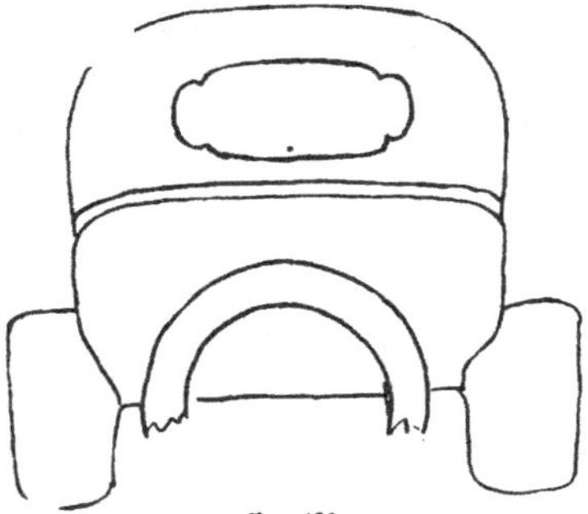

Fig. 426

Fig. 427

A collection of rear light designs arranged for the benefit of designers or students. They are all taken from designs that have been actually built. (Figs. 416 to 427 inclusive.)

doors so that the passengers can get in and out. Suitable lights made of pyralin are provided and so arranged as to give a maximum amount of vision for the passengers. There are two arrangements of side curtains covered by patents called the "jiffy" and the "Collins," both of which are used extensively. The

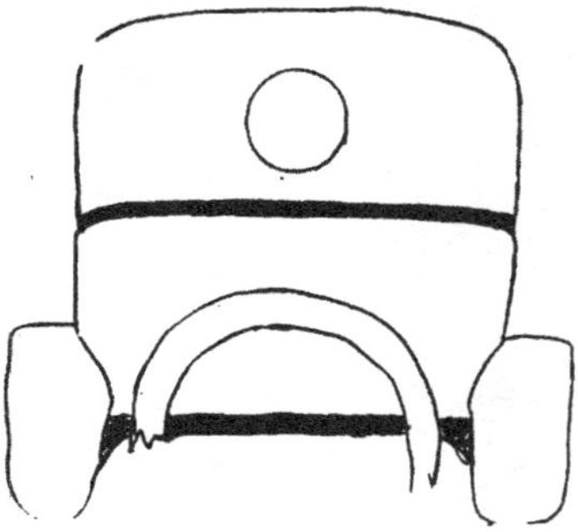

FIG. 428
A large, round glass like this does not allow either the side passengers in the rear seat or the driver to see out of it.

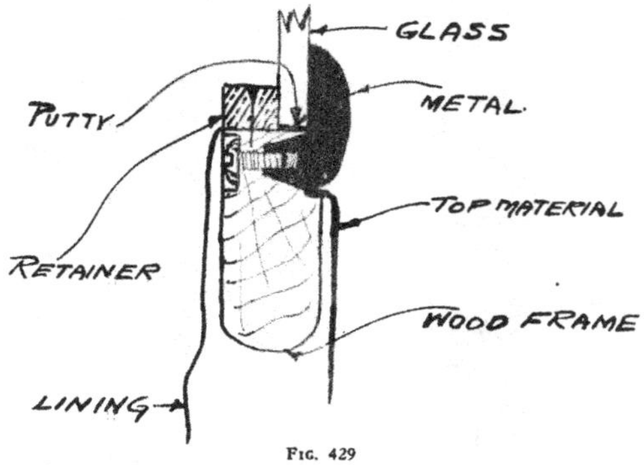

FIG. 429
Diagram showing method of installing Johnston type of light.

jiffy style of side curtain is shown in Fig. 432. The material is so sewn together that it will quickly fold up into a compact space. A cable is provided inside the top and the curtains are strung on to it, so that a movement ahead or backward folds or unfolds the curtain very easily. The curtains are usually divided into three sections on each side, and when they are not in use they fold up and fasten on to the top bows by straps, the right and left section folding together at each place. This is shown in Fig. 433.

The Collins system, as shown in Fig. 405, provides a small wire loop, for the curtain to be attached to, the idea being to keep the curtains located adjacent to the part of the top and body that they attach to. See Fig. 434. As each of the

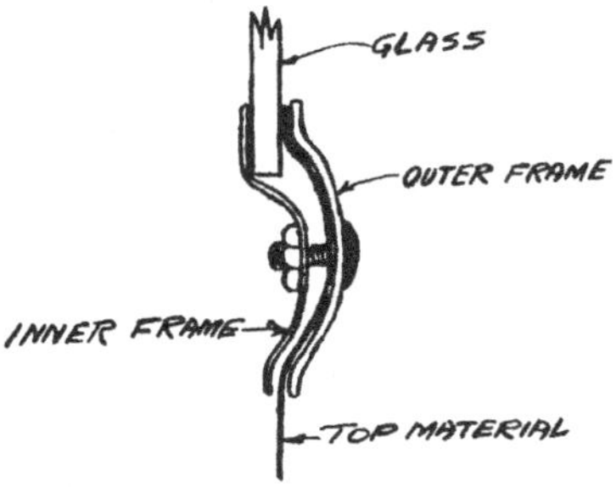

FIG. 430
Diagram showing method of installing Brewer-Titchener style of light.

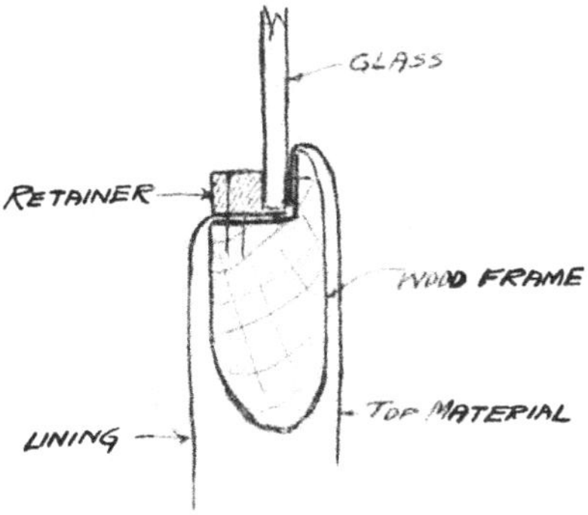

FIG. 431
Method of installing light with wood frame is shown in this diagram.

three curtains is of a different shape a lot of time is saved in putting them up when they are located by these loops and straps, as without any locating straps it takes quite a little time and trouble to sort out the curtains when it is desired to put them up.

In order to attach these curtains to the top suitable, quick detachable fasteners have to be used and a brief description of these will prove interesting. A fastener of a type widely used is shown in Figs. 435 to 439. Fig. 435 shows the fastener used when it has to be attached to the solid part of the body or top, and as it is seen is attached by two screws. This, and all the styles, are made in what is called double and single type. The single type is for use where only one thickness of cloth is to be attached to it, while the double will take two thicknesses.

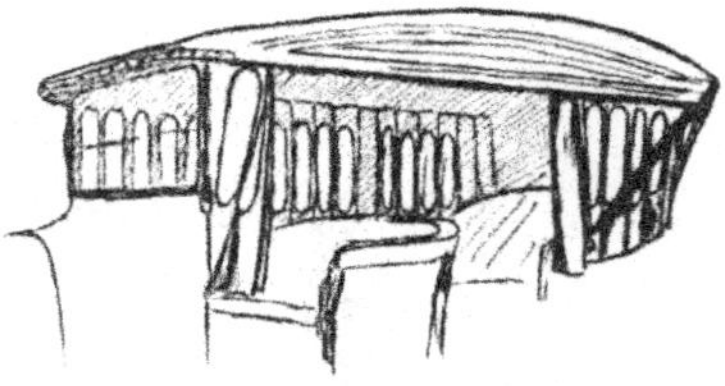

FIG. 432
"Jiffy" style of detachable side curtains.

The eyelet and plate shown in Fig. 440 are the parts that attach to the cloth. The prongs are clinched down over the plate as shown in Fig. 441. The type shown in Fig. 436 has a wood screw end on it for use when it is desired to screw it into the wood, while that shown in Fig. 439 has a metal thread for screwing on to a metal section. In Fig. 438 the fastener has a plain stud on it. This is used where it is desired to rivet it to a plate. A type to attach to cloth is shown in Fig. 439. This has two prongs which fasten over a plate as shown. The Murphy

FIG. 433
When "Jiffy" curtains are not in use they may be folded and fastened to the top bows by straps.

type of fastener is in operation at right angles to its normal position as shown in Figs. 442 and 443. Fig. 442 shows the fastener in position in an eyelet; in Fig. 443 it shows the head turned and locking it in position.

While the catch portion of the various makes of fasteners is different, the back portions that attach to the various parts are similar to constructions described in the Murphy. The "lift the dot" fastener is one used very extensively and can be adapted to practically any conditions. They are made with the different backs to suit the part they are to be used with. In Fig. 444 is shown the main portions of this type of fastener. The catch part has a small wire spring in it that snaps over the ball-shaped end of the other portion. The stud is attached to one part and the fastener to the other part that is to be fastened. It will be seen that a plate and prongs are provided to clinch over the material.

The Carr glove fastener is used a great deal on parts of the top and side curtains. These fasteners are made to attach to the cloth on wood and in metal. The principle of their construction is shown in Fig. 445. This is made of three

Fig. 434 Fig. 435 Fig. 436 Fig. 437 Fig. 438 Fig. 439

Method of attaching curtains by small wire loop, as in the Collins system. See Fig. 405
A type of curtain fastener that is widely used is shown in Figs. 435 to 439.

parts—a cap, clasp and stud. The cap has an extension that clinches over the clasp portion. The clasp has a small wire spring in it that snaps over the stud when it is engaged. The Cinch fastener is shown in Fig. 446. By depressing the small knob it permits the curtain to be removed. There are three parts, the catch, plate and washer, shown in Figs. 447, 448 and 449, respectively. This construction is very ingenious and positive in action.

Fig. 440 Fig. 441 Fig. 442 Fig. 443

Parts and operation of fasteners are described on page 194, opposite.

When the top is folded, a suitable method of holding it securely in position has to be used. The earlier method was to strap the pockets in position as shown in Fig. 450. The rest is attached to an iron which projects from the body. A top holder that is now in general use is called the Bair top holder. This consists of two main malleable iron sections with a suitable lever on the top which clamps the main section tight on to the bows and holds them securely in position. In Fig. 451 is shown this holder in position and the dotted lines show it partially open. The holder attaches to an iron which projects from the body and fits into a socket in the body iron.

The holder shown in Fig. 452 has an extension on it. This is cut on an angle as shown so by tightening the cap screw shown it will be held very securely. The body iron which is used with this type of socket is shown in Fig. 453. When it is desired not to carry the top holders on the body they may be easily removed by loosening up the cap screw. In order to make a finish on the body the cap shown in Fig. 454 is used to plug the hole. There is a spring tension in the bottom that keeps it in place. In Fig. 455 is shown a new style of front top iron. This is fastened in the body in the regular way, but instead of the top iron stud being fastened to the iron it is made detachable as shown in Fig. 456. The stud uses the same locking system as the top holder which is a splint end section. The

FIG. 444

FIG. 445

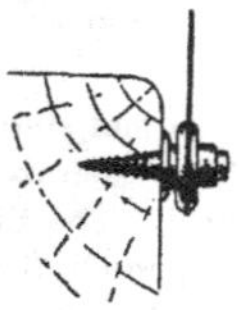

FIG. 446

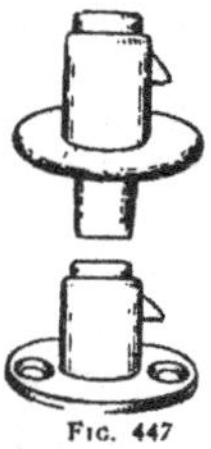

FIG. 447

FIG. 448

FIG. 449

These small drawings represent the construction and show the method of operating certain typical fasteners. Descriptive text on page 195.

advantage of using this style of iron is that the top stud can be put on after the body is painted so there are no studs projecting from the body while it is being painted and trimmed.

There are a number of different styles of holders on the market besides those described, but they are not used as extensively as the Bair type, so they will not be taken up at this time.

With the top folded it is necessary to provide a cover to keep the dust from collecting in it. The cover used is called a "top boot" and is made from a cloth similar to that used on the top covering. The boot is made to fit snugly over the sockets and extends as far as the trimmings on the seat back. In order to hold it in position two or more straps are fastened to the front upper edge and two or more to the front lower edge. These straps pull down over the seat trimming, and are secured to the seat frame under the cushion, as shown in Fig. 457. In order to make a neat appearance on the sides two extensions can be provided that run to the nut on the top iron. A fastener is located on this which helps to hold it in position.

The standard type of top has been described in most of its phases. There is another design that deserves some comments and that is the top that is constructed to fold into a compartment in the back of the body. There is (or was, at least) one manufacturer who is supplying a top of this kind as regular equip-

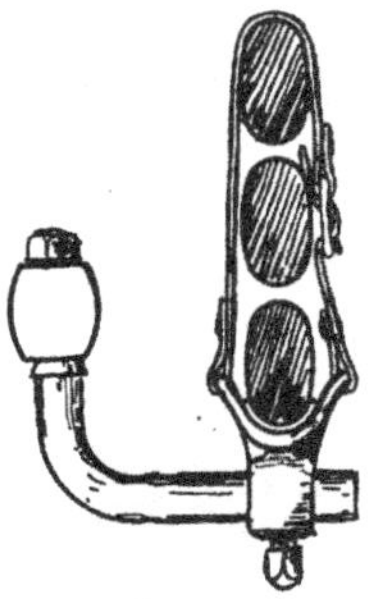

Fig. 450
An earlier form of top holder.

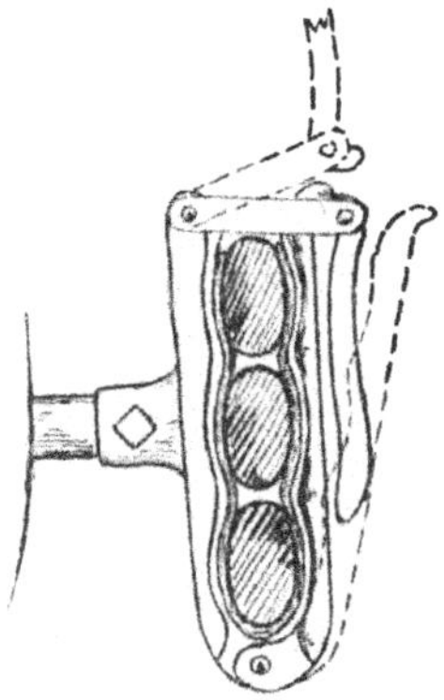

Fig. 451
The Bair top holder.

ment. In order to make a satisfactory installation of this type of top, the body has to be specially constructed, and the general design adaptable to it. In Fig. 458 is shown a car design that has them worked out to get these results. It will be noted that there is no visible evidence of the top when it is folded away, as it is on the car illustrated.

Another type of body that has a compartment for a top is shown in Fig. 459. The lines show the general construction of the top. It will be seen that the lines and shape of the top are different from the regular car top. This is caused by the conditions that have to be worked out to make the top fold close to the body and so keep the compartment from getting too large or cumbersome. The advantage of a top of this type is the protection that the top gets from dust; it also does away with cloth covers and loose flaps, and in some cases it is possible to raise and lower the top quicker as the top holders, top cover and socket holders do not have to be manipulated. The disadvantages are that the shape of the top is not as good as the regular type and that the body construction is complicated and the lines of the car are changed by the addition of the compartment.

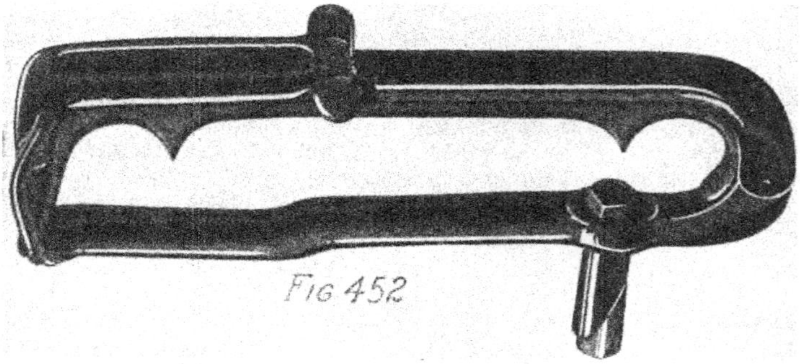

FIG 452

This holder has an extension on it, cut at an angle as shown, so that by tightening the cap screw it will be held very securely.

FIG. 453

The body iron used with the above type of socket is shown in this plate.

FIG 456

FIG 455

FIG 454

Fig. 454 is a cap used to plug the hole; Fig. 455 is a new style front top iron; Fig. 456 detachable top iron stud

The Victoria top is quite popular on cars for use in the city, as it has a striking appearance and is very adaptable for city use. The construction is identical to those tops that were used on carriages. The general form of construction is

FIG. 457
Dust cover or "top boot," in place held by straps secured to the seat frame, under the cushion.

FIG. 458
Special body with built-in compartment for disappearing top.

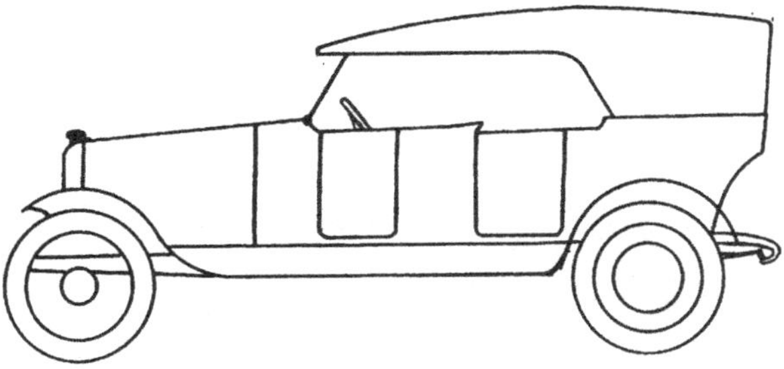

FIG. 459
Another type of body that has a compartment for a top.

to use from three to five bows with outside joint irons to hold it in position, as shown in Fig. 460. The object of the design is to have a top that affords protection from the sun, and it does not obstruct the vision as much as a regular top does. It is also possible to fold it down when an open car is desired, and then it

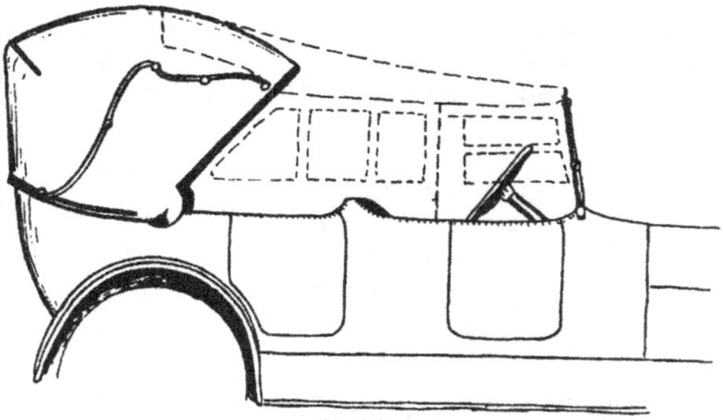

FIG. 460
The general form of construction of the Victoria top is to have from three to five bows with outside joints.

FIG. 461
A solid construction, eliminating the joint iron. This top cannot be folded.

FIG. 462
A car with detachable winter top, which can be fitted very closely.

has the appearance of a regular folded top. The best grade of leather is generally used to cover the top. It is not advisable to fold it down, as the leather will crack or show the line of the fold, and for this reason these tops are not folded very often.

To afford proper protection in case of rain an extension curtain is provided that will attach to the windshield, and then provision is made for side curtains. This is shown in Fig. 460.

Fig. 463

Fig. 464

Fig. 465
The majority of the early detachable tops were crudely constructed, but from these early attempts there has been evolved a top that is proving satisfactory, as shown in Figs. 463, 464 and 465.

There are two methods of building these tops. One is what is termed a shifting rail and the other is where the top is built so that the covering fastens to the body. The shifting rail type has a rail that fastens to the body, the top covering being fastened to it. This permits the top being taken right off the body. When this shifting rail is used the rail makes the top project over the sides of the body. This has a tendency to make the top appear larger, as shown in the illustration in Fig. 460. For this reason it does not look as well as the top that is built on to the body. The increasing popularity of the Victoria top has caused manufacturers to devise a top construction that is solid, eliminating the joint iron so that it cannot be folded, and then setting a fancy shaped glass in the sides as shown in Fig. 461. This gives a Victoria effect, but with smooth lines and simpler construction.

The winter top is a construction that has been developed to afford more protection to passengers in the winter, and as the automobile is being used, to a great extent, all the year round, the detachable winter top construction is being used quite extensively. The winter top virtually converts a touring car into a closed car. There are a number of well-constructed tops on the market that make a very convenient adaptation for winter use.

This detachable top, seen from beneath in the illustration, consists of a wood framework with composition or metal panels, sides and roof. The composition panel is generally used and this is covered with a good grade of leather substitute.

One manufacturer has made a specialty of a detachable top on his cars for several years. The body is especially arranged so that the top can be fitted very easily. This top and the details of the attachment are shown in Fig. 462. A great variety of winter tops have been marketed of various designs. Unfortunately the majority of them were very crudely constructed, and they did a great deal towards retarding the development of this business. From these early attempts there has evolved a top that is now proving satisfactory and acceptable. See Figs. 463, 464 and 465.

A brief description of a type of top now being used will be given. The top consists of a wood framework with composition or metal panels, sides and roof as shown in Fig. 466. The composition panel is generally used, and this is covered with a good grade of imitation leather. The top, back and side quarters are

The upper sections of the doors on the detachable top are attached to the regular car door as shown in this photograph.

A detachable top assembled on a car body, ready for painting. The glasses in the doors are constructed so that the upper half can be lowered when desired.

attached together and fastened to the windshield at the front, and to the rear top iron at the rear. The rear side glasses are set in a permanent frame which is attached to the top and body by small steel plates. The center posts from which the upper part of the door swings are fastened to the body and top by suitable steel plates. The upper sections of the doors are attached to the top of the car door as shown in Fig. 467. A handle is provided in this upper section to operate the door from the outside. This has an extension on it on the inside that attaches to the regular door lock handle. This device is necessary, as the majority of automobiles have the door lock handles on the inside of the doors.

The glasses in the doors are constructed so that the upper half can be lowered when desired, as shown in Fig. 464. On some of the winter tops they are arranged so that the upper door frames, side posts and rear quarters can be removed for summer use, giving a "Springfield" effect, and to provide against sudden storms a set of side curtains are fitted. These fold up and set in a pocket in the top lining. Fig. 468 shows a top assembled on to a car ready for the paint.

The permanent top construction has not proven as popular as was expected. One of the reasons was the poor appearance the first designs gave to the car. This was sometimes caused by the top builder not making the top the right proportion. It was found very difficult to adapt a permanent top to many of the open bodies, as their contours or lines and construction did not lend themselves to a permanent top installation. There is a gap between the open body and the closed which led to the development of the so-called "California top" design. These tops can be really classed as permanent ones, but they were built to follow along the lines of the regular collapsible top, though elaborated upon considerably. The succeeding chapter describes the latest tops.

THE CALIFORNIA TOP

The California top is a recent development which originated and developed in the West. It is really an elaborated cloth covered top either with a permanent non-folding frame or a folding frame. The majority of tops use a solid frame, as it permits making better outlines and forms a better support for the glass lights. A fancy grained or colored covering is used and this combined with nickeled trimmings and plate glass gives a very novel and distinctive appearance. The real beauty of these tops is not brought out by photographs, as the general finish and color all play an important part in making these types of tops pleasing and popular. The photographs Figs. 469, 470 and 470-A show three characteristic California tops.

The first consideration in laying out a top is to get a side view or what the designer calls the "side elevation," as this is really the most important part of the California type top. The contour and arrangement of the side determines the whole design of the top. Fundamentally the roof determines the side view, and the roof is proportioned to extend from the windshield to the rear of the body, and must be wide enough to cover the sides of the body. Its height from the body must be sufficient to allow ample head room for the passengers. This is a point which is often overlooked by designers, who are intent on getting a definite outline.

A fine appearance is always desired, but if it sacrifices the comfort of the passengers it is of no practical value, and instead of being a product which has a high market value it will be a "white elephant." Should it be sold and the owner hits his head and is otherwise made uncomfortable, there will be a dissatisfied customer who will do a lot of harm to the business.

To illustrate what is meant, the drawing shown in Fig. 471 gives a typical five-passenger body with a California type top on it. On the front seat there are 37 inches head room or clearance between the cushion and the lowest projection on the top, which is the cross bow. About 1 to 1½ inches can be figured on for deflection on the cushion so there are actually about 38½ inches for head room or clearance. For the rear seat there are 38 inches, and with the cushion depression added, about 39½ inches for the passenger clearance. In Fig. 472 is shown the same body with a fancy design of top on it. The designer has cut the back corner off and also lowered the front a little to get a rounding effect, and at the same time he has not raised the sides at all, but retains the overall height. The result is that all the headroom he has for the front seat is about 36 inches, and about

A California top, designed on good lines and furnished with side curtains. No framework is visible around the glass lights. The covering material is leather substitute in a color to harmonize with the car painting.

California top with metal frame glasses. Window glass is usually beveled-edge plate and is furnished assembled in the frames ready to use by several manufacturers.

36 inches for the rear seat. This is entirely too little and the man who is a trifle above the average height cannot wear a felt or derby hat. The first bad bump the car encounters he will hit his head.

In laying out a top the outside, or top lines, are to be determined at first. After this is done the width, or sides, should be worked out. This is not quite as simple as it seems at first glance, and it will require considerable judgment to obtain good results.

Every make of car has peculiarities which must be considered separately. For instance, sketch Fig. 473 shows the front end of a typical open touring car, and it will be noted that the front of the top is considerably wider than the wind-

FIG. 470 A

A California built car with top of striking design. This top is not built on to the car body, but with it, and is therefore stronger and more rigid than the usual California top. (Leach-Biltwell, Los Angeles, Calif.)

shield, as indicated by X-X; one of the first thoughts will be to cut this down. Before this is done the relative widths of the windshield and the body must be considered. If the body and shield are wide it is possible to make the top narrow at the front; but if the body is narrow care must be exercised. If the top is made too narrow the driver will not have proper clearance for his arms when the side curtains are up.

The width of the windshield is governed by the width of the body at the front, while the width of the top is determined by the width of the body at the rear, as it has to be wide enough to clear the body when the top is folded down. This latter condition does not, of course, apply to the California top. Fig. 474 shows the average width of top compared with lines of the body.

Coming back to the side elevation, there are the varying heights of the sides of the body to consider. In sketch Fig. 475 is shown a top fitted to a low sided body. It is apparent that head room has to be provided disregarding the height of the side of the body. Consequently there is a larger opening between the sides

THE "ROAMER" CALIFORNIA TOP, AS PRODUCED BY THE BARLEY MOTOR CAR CO., KALAMAZOO, MICH.

The different views show the winter panels, the summer curtains and a side and rear view of the fully opened top.

of the body and the quarters of the top, making a large side curtain necessary. In Fig. 476 is shown the same shape of top, but the body sides are higher and consequently we get much better balance in the design with a corresponding improvement in the appearance.

In an ordinary folding top the sides of the back curtains are generally extended to make what is termed a gypsy curtain, as shown in Fig. 477. With the

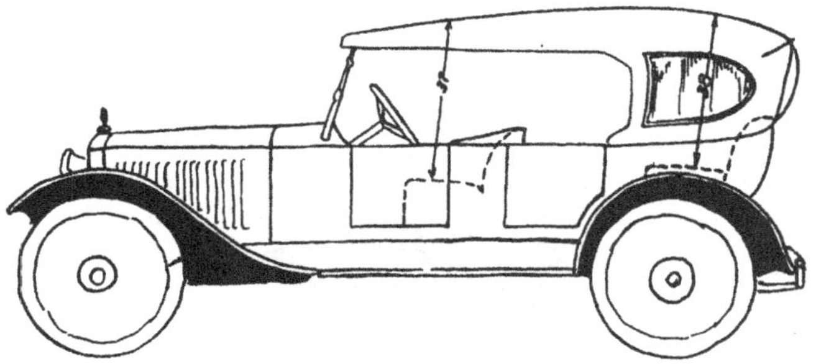

FIG. 471
Typical five-passenger body with a California type top. This design gives ample head room.

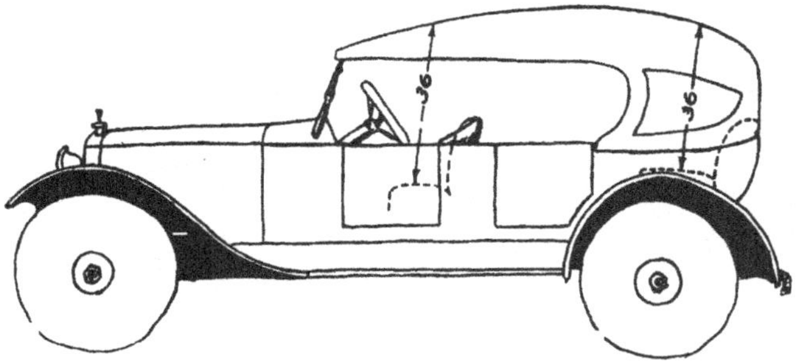

FIG. 472
Same body as shown in Fig. 471, but with a top of more "fancy" style. The rounded back cuts down the rear seat head room to about 36 inches on the rear seat. The rounded front also reduces the front seat head room to the same number of inches.

California type the entire space from the back of the top to the rear door may be used for support and frame for side lights. This side curtain offers, perhaps, the best chance of all to designers of California tops to display their ability in design. The skillful arrangement of the side lines of the quarters and back support, combined with the properly proportioned side or rear light, will accomplish wonders in making a California top right.

It will be quite interesting to the designers to examine a number of side elevations or views, of California tops which have been built, so that they may start their design with a knowledge of what has already been done. It is always desirable to be original, so for the designer's reference a number of the most striking designs have been collected and outlined in sketches 478 to 484.

The body and top designer has always to contend with the general construction of the car, its size and wheelbase, when laying out the bodies and tops. When

FIG. 473
On this top, note that the front of the top is considerably wider than the windshield, as indicated by X on either side.

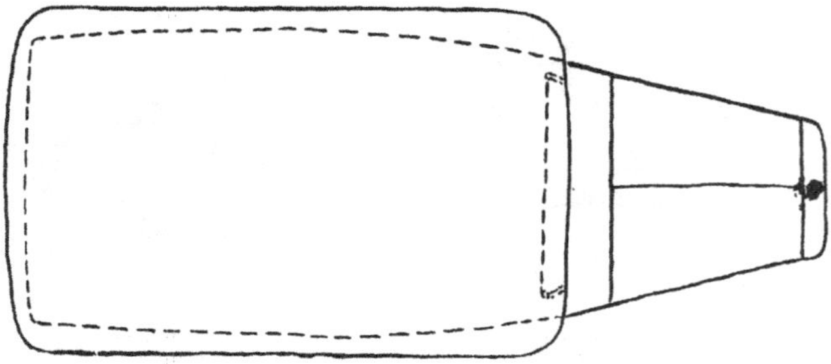

FIG. 474
This drawing shows the average width of the top as compared with the lines of the body.

the top is laid out, the length and type of body play a very important part, and consequently a top that would look fine on a large car may not do at all on a smaller car, and *vice versa*.

The top shown in Fig. 478 is what we would call a rounded effect, and has a large side light. The rear end has a radius and the front end is curved in with a radius similar to the rear. The glass is set in a frame which can be taken com-

FIG. 475
California top fitted to a low sided body.

FIG. 476
Here, the body sides are higher than in Fig. 475, consequently we get better balance in the design.

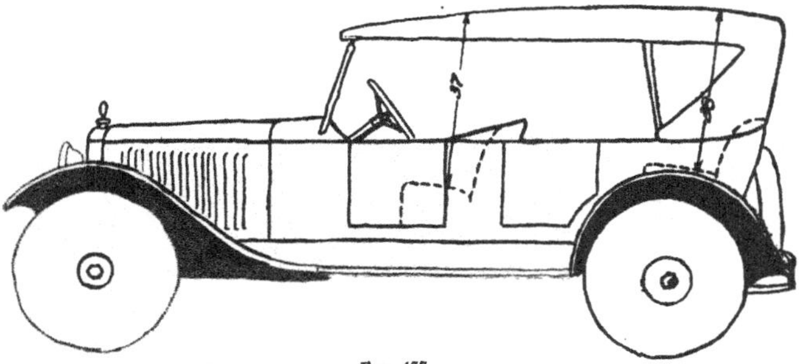

FIG. 477
In an ordinary folding top the sides of the back curtains are generally extended to make what is termed a "gypsy" curtain.

Fig. 478
This top has a rounded effect and has a large side light.

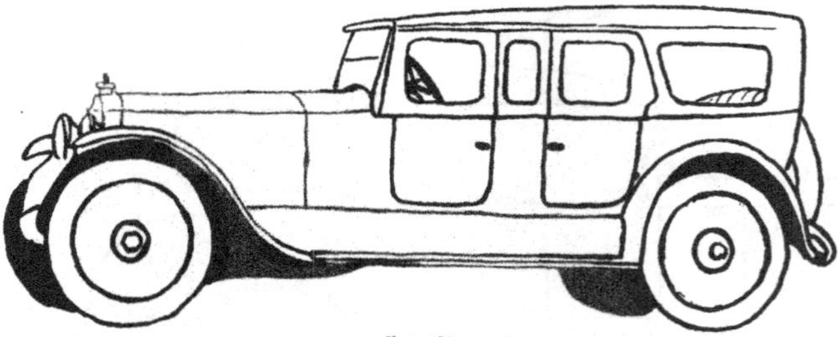

Fig. 479
All the lines on this top tend to give a straight effect, approximating quite closely the average closed body lines.

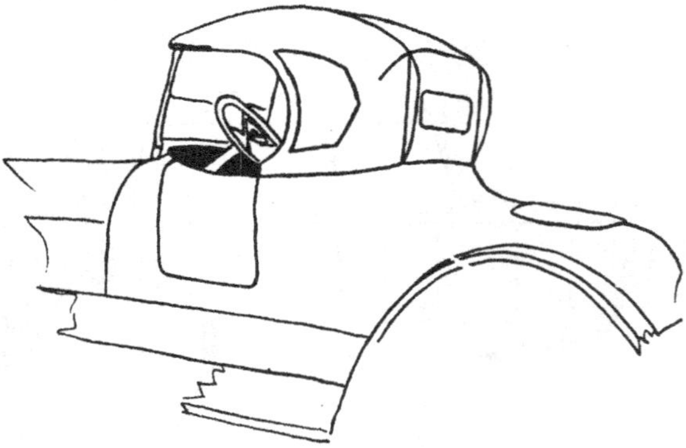

Fig. 480
A roadster type, which has a solid side frame and a rounded roof, with a fancy shaped side glass.

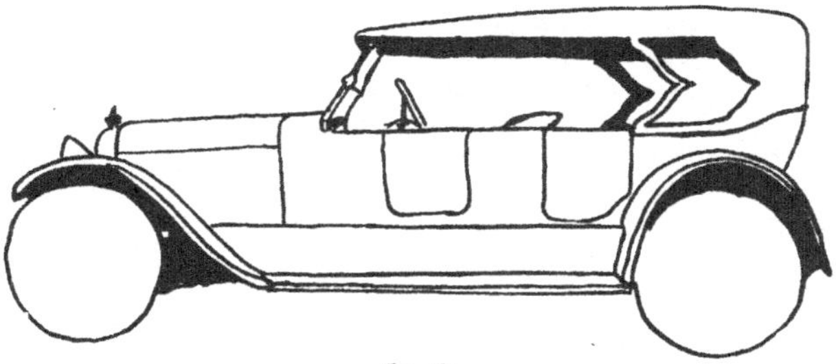

Fig. 481
This top has comparatively straight lines with a solid side and an unusually elaborate side glass.

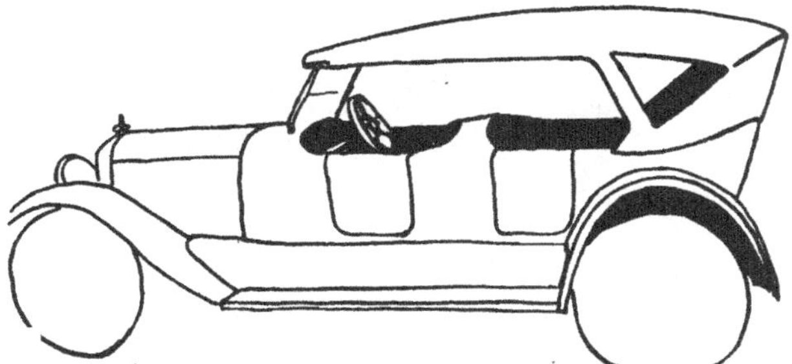

Fig. 482
California top, which looks much like the ordinary collapsible top, with the exception of the large triangular side light.

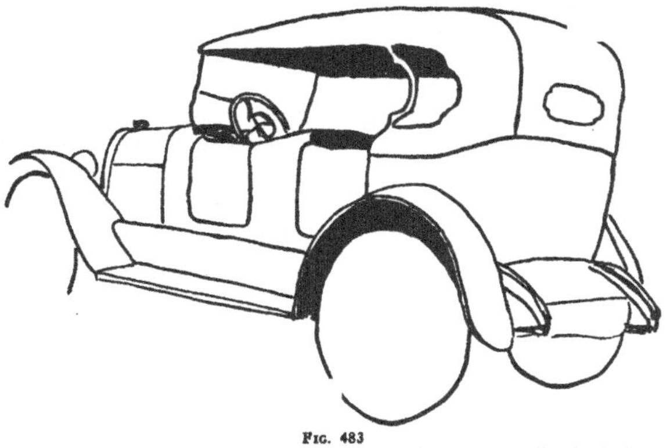

Fig. 483
A rounded type of top with a fancy shaped side and side light. The back light is also rather ornate in design.

pletely out of the sides of the top if it is required. This means that the rear end of the top must be strong enough to support the roof with the glass removed. The roof has a well rounded front and rear end and so harmonizes with the side lights. In Fig. 479 the roof of the top is partially supported by the side construction and so differs quite radically from the one shown in Fig. 478. All the lines on the top shown in Fig. 479 tend to give a straight effect, approximating quite closely the average closed body lines.

Fig. 484
Victoria type of top made on a solid frame. This is a "smart-looking" job and can be built more easily than a regular Victoria top.

A roadster type is shown in Fig. 480, which has a solid side frame and a rounded roof, with a fancy shaped side glass. The top shown in Fig. 481 has comparatively straight lines, with a solid side and a very elaborate side glass. In sketch 482 is outlined a top which compares very much to an ordinary collapsible top with the exception of the large triangular side light. The side on the top is solid.

A rounded type of top is shown in Fig. 483 with a fancy shaped side and side light. A Victoria type of top made on a solid frame is shown in Fig. 484. This makes a very smart looking job and can be built much easier than a regular Victoria top. Dummy hinges are shown, but these may be omitted if desired.

CHAPTER XIX

TRIMMING OR UPHOLSTERING

The interior finishing and upholstering of the automobile body is generally the last consideration in the car. It is, however, the first consideration of the passengers. A perfectly constructed car would not be acceptable without comfortable upholstery. Modern cars are expected to act efficiently and the deciding factor of a sale is often the cushion and trimming arrangement. Comfort or ridability is the first consideration of the upholsterer and next comes art in designing and the selection of the materials used.

Typical example of trimming on a seven-passenger touring car or "phaeton."

Upholstering is a trade dating back several centuries and the automobile takes from the experience of this trade and adapts whatever it requires to its use. The automobile trade applies the term "trimming" to the art of covering and arranging the car's cushions and appurtenances, so the term "trimming" will be used throughout this chapter.

Two distinct divisions can be made in trimming. These are open body and closed body trimming. Each has its separate problems but uses the same principles

Fig. 486
Typical example of trimming on a roadster body.

Fig. 487
Door trimming and dash equipment on same car as in Fig. 486.

Example of rear compartment or tonneau trimming, showing door and back of front seat.

Another method of treating the door and back of front seat in a touring car or "phaeton."

throughout the main part of the work. The open bodies do not require as much work to trim as the closed because there is less surface to cover. The shapes of the cushions and backs are very similar and the springs are practically the same. The open bodies use leather or imitation leather throughout, whereas closed bodies use a large variety of covering materials. Figs. 485 to 489 show typical

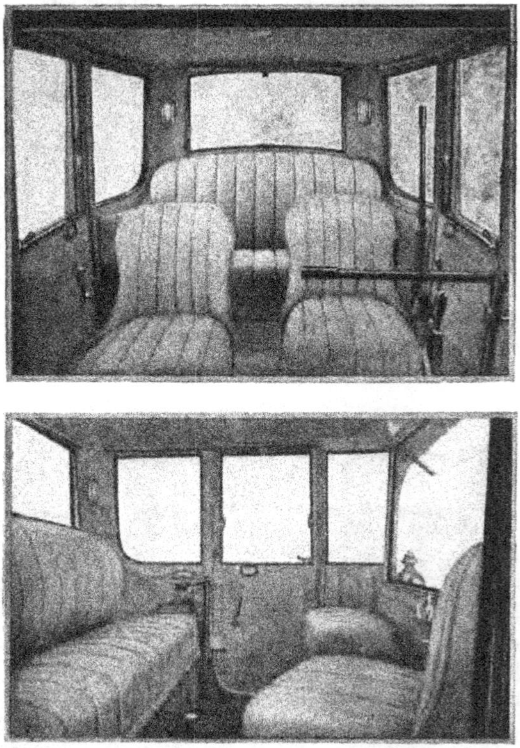

FIG. 490
These two views give a good idea of French plait upholstery as applied to a well-known electric car.

examples of open car trimming. A seven-passenger touring car is shown in Fig. 485 and a roadster in Fig. 486. Closed car trimming is well illustrated in Figs. 490 to 493. These will be taken up in detail further on.

The materials used for trimming cover a wide field of manufacture and deserve very careful attention, especially in large production work, because they are expensive. Some of the materials used will be briefly described. Open car trimming uses leather and imitation leather for the main covering.

Leather as used for trimming, can be obtained in a variety of grades and grains. Hand buffed and machined buffed in long or pebble grains are used mostly. Specially treated hides in all colors and grains are obtainable for special cars. Brown morocco is perhaps the most popular among the special grades.

For economical trimming imitation leather can be used. Imitation leather is one of the modern developments and it is only within recent years that imita-

FIG. 491
A pleasing example of French plait upholstery in a gasoline car.

tion leathers could be made which closely imitate the grain of leather and give satisfaction in appearance and durability. While they are very good they do not quite compare with the genuine leather for lasting service and durability, consequently they are mostly used in places where they are not subject to extraordinary wear, such as door and door pillar, covering rear of seats, etc. Closed cars very seldom use leather except for the driver's seat in limousines. The materials used in closed cars depend on the style, bedford cloth, cotton worsted cloth, tapestry cloths, mohair, velour, cotton velour, velvets and silk, are all used when desired.

The body is of course, the foundation that the trimmer has to work to, but the most important thing for the trimmer to do is to arrange the back and cushion springs. A brief consideration of the springs will be of interest.

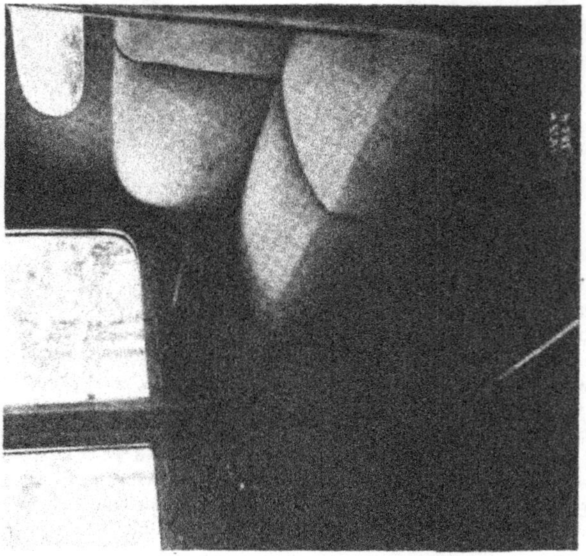

FIG. 492

Plain, smooth cushions are hard to make properly. They have a tendency to show wrinkles as soon as the hair packs down a little.

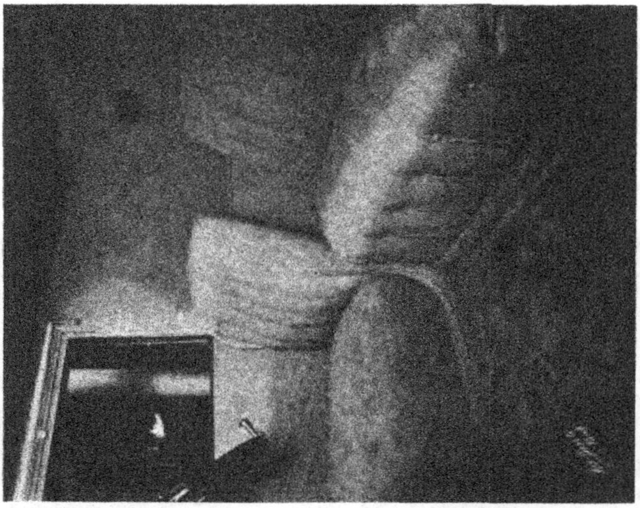

FIG. 493

One-button plait upholstery in a Minerva car at the 1921 New York Salon. (Smith-Springfield Body Corporation, New York)

SEAT CUSHION SPRINGS

Modern car springs are arranged to give a long, easy action, or expressing it in another way, the nodes of vibration are long, and this means when a car meets a rut or lump in the road the passenger load will be deflected with a long, easy motion and not the "choppy" action which was formerly so unpleasantly familiar to users of motor cars.

Cushion springs are provided to prevent the passengers getting jolts when the chassis springs act, and they also absorb vibrations which the stiff chassis

FIG. 494

FIG. 495

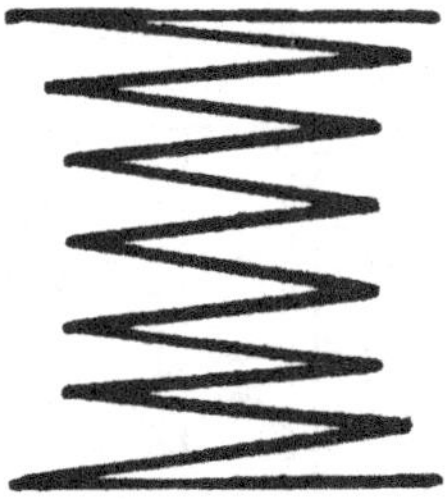

FIG. 496

FIG. 497

In Figs. 494 to 496 are illustrated the shapes of the cushion springs referred to on page 225. In Fig. 497, is shown a section through a typical "double-deck" spring construction.

springs won't absorb. There is a wide range of opinion regarding the most desirable qualities a cushion should have. An easy-riding cushion, to the majority of people, is a very soft cushion or a cushion that feels very soft when it is sat upon.

If the cushion is too soft it is very uncomfortable in the summer time; if it is too hard it does not have the luxurious feeling which seems to be so desirable. However, if the cushion is too soft a sharp bump on the road will allow the springs to collapse completely and the passengers will go through and hit the seat bottom.

Cushion spring constructions, as the frame works are called, use coiled springs in some form or other. This means that the passenger load is supported by a compressed coil spring and when sudden jolts are encountered the spring is further compressed; but as soon as the shock is over they tend to return to normal, and if their action is not retarded in some way the passenger is raised

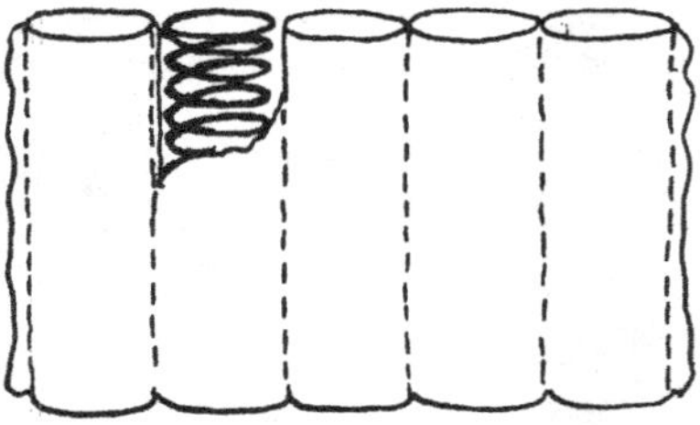

FIG. 498
The "Marshall" type springs are each enclosed in a cotton cylinder, the springs being packed closely together.

up or thrown very suddenly and with all probability of hitting the roof of the car. The various types of cushion springs which are manufactured and used will be analyzed so it will be possible to have a good working basis for the development of designs. It will be necessary to pass over the many small variations in the

FIG. 499
A typical spring assembly with U-section metal base.

spring design in order to make a suitable grouping of the main principles of construction.

The types will be classified into five groups; first, single cone; second, double cone; third, cylinder; fourth, double deck; and fifth, the Marshall spring. The first type, the cone or "single cone," as it is called, is used a good deal on omnibus cushions and shallow cushions of all types. The object of making the spring cone-shaped is to make it stiffer or to give it more resistance as it is compressed. This spring is suitable for shallow seats and for vehicles that do not have a high range of speed. Its average height is 4 inches.

FIG. 498 A

FIG. 499 A

FIG. 500 A

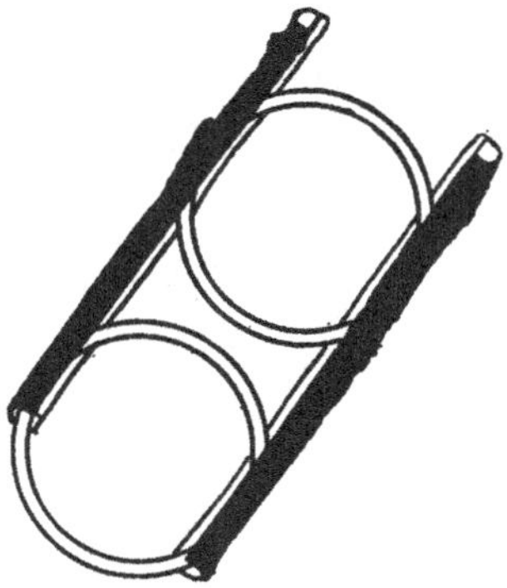

FIG. 501

The springs are clipped into the base as shown in the enlarged section, Fig. 498-A. Another method is to have a section like that shown in Fig. 499-A. The cushion cover is held on the spring by tacking the lower edges, as in Fig. 500-A. The outer spring clips to the inner edges of these bases, the other springs are held in place by V-strips as shown in Fig. 501.

For the ordinary passenger car which has to travel a wide range of speed it is necessary to supply a deep spring so it will have an opportunity to absorb the extreme shocks that it is subjected to. The average height for this type is 6 inches. The gauge wire used on both single and double cone varies from No.

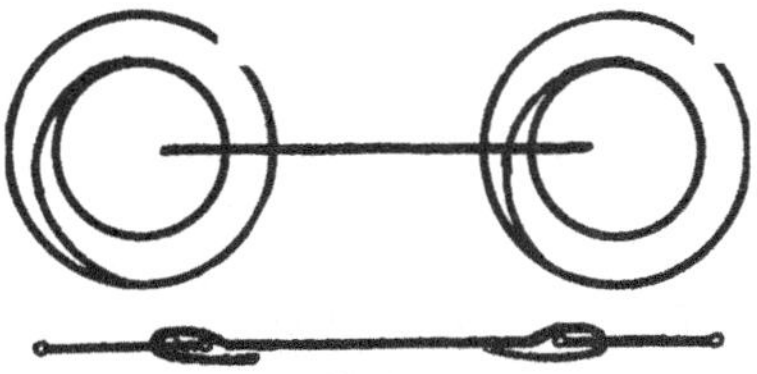

FIG. 502

The centers of the springs are tied together with wire or clips similar to the manner shown in this diagram.

10½ to No. 9 (old English or Birmingham wire gauge), number 10 being used the most. The cylinder type spring has its center portion wound parallel but has the upper and lower coils enlarged to make a satisfactory attachment to the spring frame. The wire sizes used in the cylinder springs are similar to those used on the double cone.

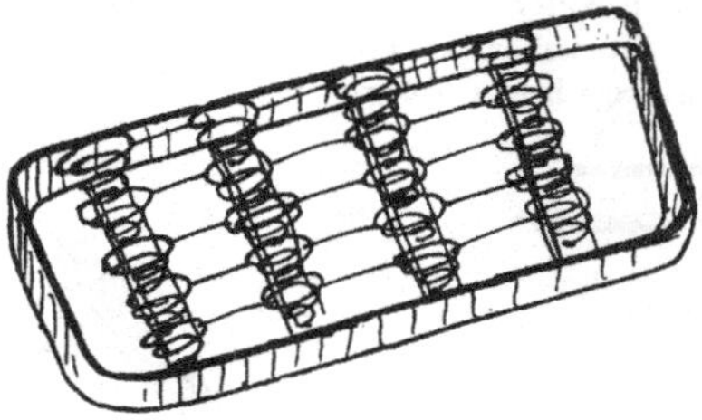

Fig. 503

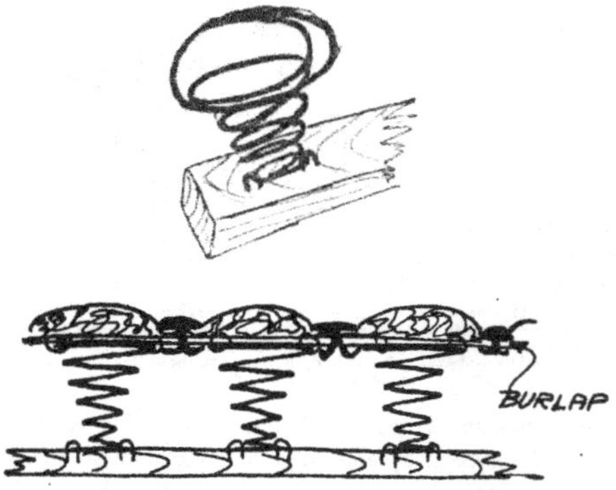

Fig. 504

Fig. 503 and Fig. 504, show two spring constructions that are quite generally used. The explanation will be found on page 228.

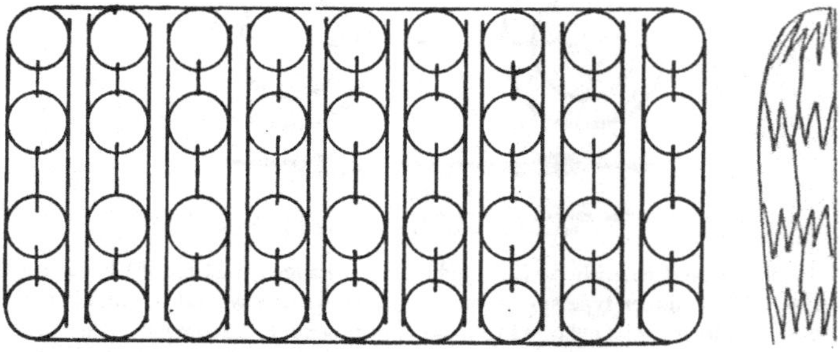

Fig. 505

Typical back cushion construction of double cone springs. This is assembled in the body and secured by leather or metal clips, as indicated in Fig. 506, on opposite page.

The cylinder type spring is designed to give a uniformly easy action when under load, while the double cone spring stiffens up quickly, having a tendency to give the throwing action that is so undesirable. The coil size on the cone spring runs from 2½ inches at the center to 3¼ inches at the largest diameter, and on the cylinder spring it runs from about 2¾ inches in the center with 3¼ inches

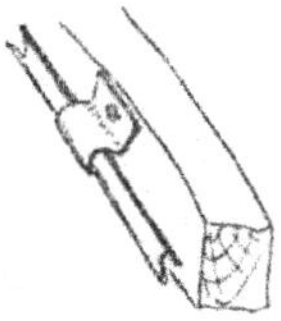

Fig. 506

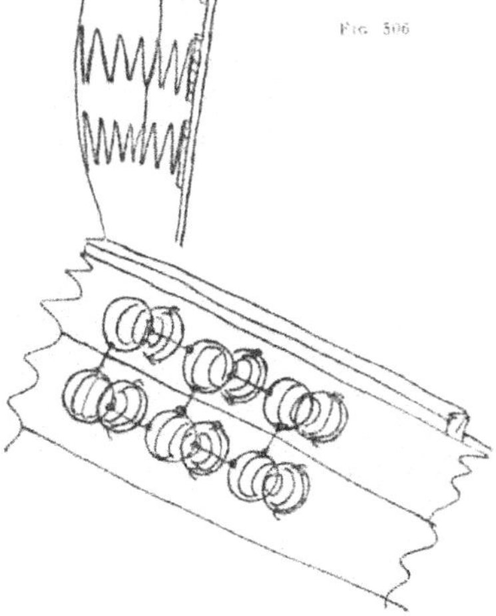

Fig. 507

If a unit spring construction is not used, each spring must fasten to the slat by staples, and they are tied together at the front with cords.

for the outer coil. In sketches 494, 495 and 496 are shown the shapes of the springs referred to. Double deck spring constructions have been developed to enable the cushion to be made very luxurious and soft. This result is obtained by having two sets of springs, one on top of the other; hence the name of "double deck." Double cone springs are generally used and the upper springs are very light, about No. 12 gauge, while the lower springs are made from No. 10½ to No. 9 gauge.

In sketch No. 497 is shown a section through a typical double deck construction. It will be noted that the total height of both sets of springs must equal the height of the ordinary spring construction. Marshall springs are very popular and they have been used for years on furniture, but their adaptation to automobiles is comparatively recent. They consist of a large number of small cylinder springs

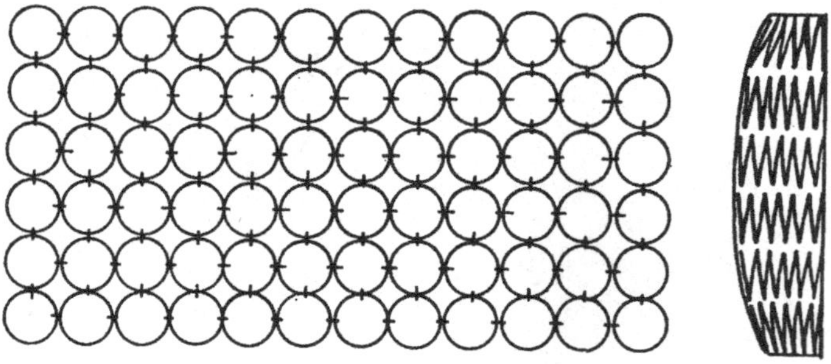

Fig. 508
The modern type of cylindrical-shaped spring has to be spaced close together to give proper supporting surface for the hair.

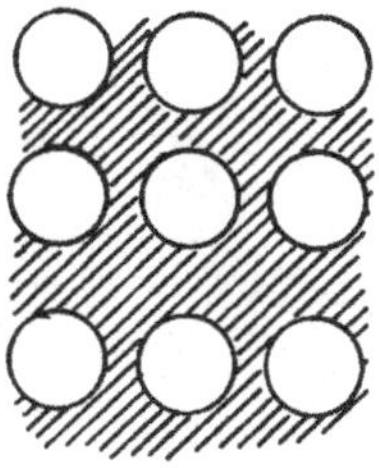

Fig. 509
Cloth laid upon double cone springs showing area in shaded lines where there is no support on spring.

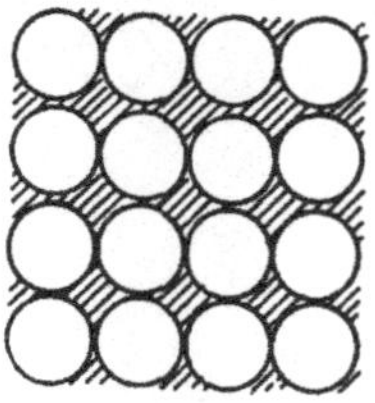

Fig. 510
Note that the shaded areas when the cylindrical springs are used are much less.

packed together. Each spring is enclosed in a cotton bag so that they are prevented from rubbing against one another and catching in the coils. This type of spring gives a very easy-riding cushion when it is made up properly. In Fig. 498 is shown a section of this type of spring. The finished cushion heights are practically the same as those previously described. Having described the main types of springs, the methods used to fasten these will be reviewed.

PRINCIPLES OF CUSHION SPRING CONSTRUCTION

The size of spring constructions varies according to the size and shape of the seats they have to fit into. Front cushions are generally narrower and shorter than the rear, and for the ordinary type of springs such as described in classes 1 to 4 there are generally about 36 springs, arranged in four rows of nine in a row. In the Marshall springs about double the amount of springs are used. For the rear cushions about forty-two springs are used, as this is generally a much larger cushion than the front, and the springs in a Marshall are increased accordingly.

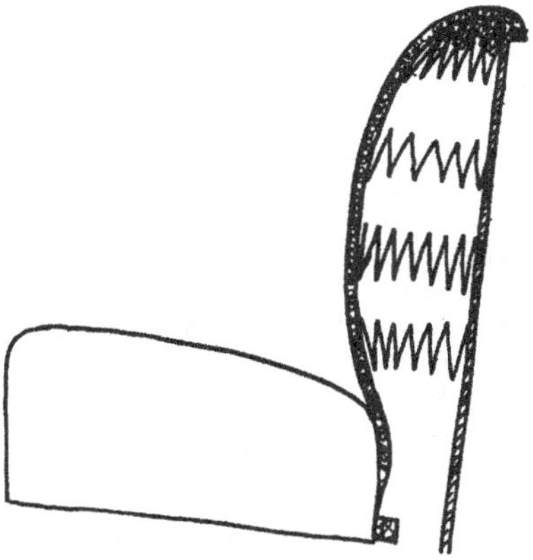

FIG. 511
A sectional view of a trimmed back.

In order to hold the seat springs together various methods are used. Wire frames or U-section steel strips are probably the most used methods, and then comes the wood base frame. In Fig. 499 is shown a typical spring construction assembly consisting of U-section metal base. The springs are clipped into the base as shown in the enlarged section in Fig. 498-A. It will be noted that the outer edge of the frame is open, to allow for the end of the cushion facing, which is clamped between the edges.

Another method is to have a section like that shown in Fig. 499-A. This is made of sheet steel, but it has a wood insertion that is just large enough to hold tacks. The cushion cover is held on the spring by tacking the lower edges as shown in Fig. 500-A. The outer spring clips to the inner edges of these bases; the other springs are held in place by U-strips as shown in Fig. 501. The upper section of the spring is made in a similar manner to the bottom, only it is not necessary to make any provision for holding the covering on the top. The covering can be likened to a bag, this is pulled over the spring and clamped at the bottom only. In order to further support the spring, various kinds of brace wires are inserted; for an example refer to Fig. 501. The centers of the springs are also tied together with wires or clips similar to that shown in Fig. 502. The braces and clips are arranged to equalize the strains imposed upon the springs and make as many springs as possible help carry the load.

Some spring constructions have a wood base or frame instead of the metal construction and still have brace wires and ties similar to those described. On the old types of construction a wood frame was used and the springs fastened to the base with staples, while the tops of the springs were tied together with cords. Sometimes a burlap cover was used and the springs were fastened to it with cord. These two methods are shown in sketches 503 and 504 respectively. The detail arrangements shown apply especially to the methods used to hold the springs in position. Now covers are also used on top of the springs to hold the trimming up and also to make the cushion construction stronger. A covering of mattress wire is used very effectively as it combines utility with economy, as it is possible to make the pads on the springs lighter without detracting from the riding qualities of the cushion.

Cushion spring coverings can be made in a variety of ways; the old method which is generally used with the wood frame that has the springs tied to cover, is to lay on the burlap some picked hair from 4 inches to 6 inches high and then the leather or substitute leather cover is pulled down over this. If the biscuit type of top is used the cover will be tied to the burlap with tufting buttons.

Biscuit type tops for moderated priced cars are made in presses. The covering material is laid on a wood form with all the buttons set through it; loose hair is then laid on this to the proper height and a piece of burlap on top. The press is operated by compressed air or hydraulic power, and it is brought down till it compresses the hair so that the buttons can be clinched over. After this operation the press is released and the top is ready to be assembled to the spring. This top is tacked or clinched to the spring according to the type that is being used.

With the advent of the plaited type of trimming a different line-up was devised. A burlap foundation was used on which to sew the plaited leather, and this gives a series of plaits that have to be filled with hair or cotton batting. When these plaits are filled we have a condition similar to the pressed hair top, inasmuch as it can be assembled over the cushion spring. If a fabric or mattress wire top is used, this stuffed, plaited top can be stretched over the cushion without any other padding, but when a plain spring is used a pad of some kind is desirable. This pad can be made from interlaced hair or mixed cow hair, according to the class of job required.

For facing or skirts, short pieces of the covering material can be sewn together to match the width or plait on top. Cotton batting can be used to give some fullness to these plaits and the backing of burlap or duck can be sewn all together. For the back and parts of the sides a good grade of heavy imitation leather can be used.

The foregoing description covers the plain or auto type of cushion. For the Turkish cushion a similar procedure can be taken, with the exception that the plaits must be carried continuously from the top to the bottom. At the front a roll must be provided on the front edge of the cushion. The Turkish cushion is luxurious in appearance, but is obviously more expensive and harder to produce in large quantities. It is probably better adapted for custom work.

BACK CUSHION SPRINGS

The trimming of the modern automobile seat back has for its foundation a combination of springs which are either made up in a frame or conveniently attached to the framework of the body. Luxury is not the prime reason for this, as it is necessary for the comfort to provide a suitable shock absorber. A fast-

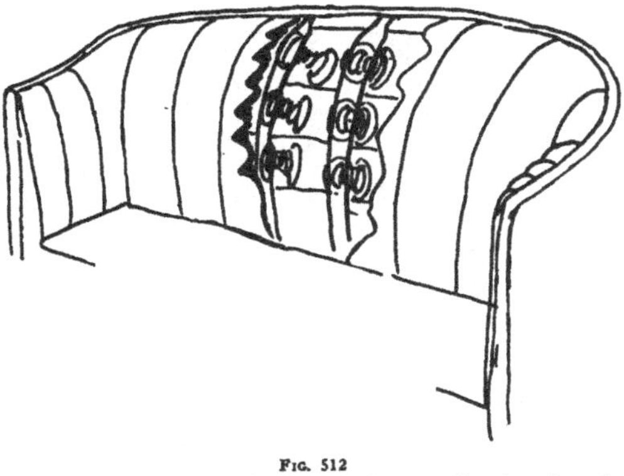

FIG. 512
In order to economize in the use of hair, several constructions have been developed, one of which is shown here and another in Fig. 513.

moving vehicle coming in contact with an uneven surface in the road would give the passengers a very severe jolt if suitable cushioning were not provided. The foregoing is meant to show the reason for the use of springs in the back of an automobile seat.

The depth or thickness of the rear cushion, in most cases, is limited by the amount of space there is in the body for the seating arrangements. In most cases the driver's seat is thinned down so that the door clearance can be made as large as possible. Then, again, the rear passengers are given the benefit of greater luxury as the ladies and older people generally ride there. On chauffeur-driven cars the rear passengers are given all the room that is possible, and the front seat back is cut down accordingly.

Back springs can be grouped into two types; the first, double cone springs, and the second the "Marshall" cylindrical spring. For convenience of handling, all the springs, except the "Marshall" type, are arranged in a frame of wire or strip steel, making an assembly called a spring construction. In Fig. 505 is shown a typical back cushion construction of double cone springs. This is assembled in the body and secured by leather or metal clips as indicated in Fig. 506. Cross bars of wood called slats are arranged in the body conveniently, behind the rows of springs, to support them. If a unit spring construction is not used each spring has to fasten to the slat by staples and they are tied together at the front with cords as shown in Fig. 507.

The cushion back construction shown in Fig. 505 has double cone springs secured at the bottom to U-shaped steel strips which are fastened to the border wires. At the top the border wire is fastened to springs which are clipped together. Clips are also provided in the center of the springs to help distribute the strains, and to prevent the springs bending up under load.

FIG. 513
In order to economize in the use of hair, several constructions have been developed, one of which is shown here. Another construction is illustrated in Fig. 512.

The double cone spring construction types just described are the cheapest to buy, as the bell-shaped end expands out so it fills up the space in the back, forming a support for the hair without using up a lot of spring wire. The modern type of cylindrical-shaped spring has to be spaced close together to give proper supporting surface for the hair. This condition is illustrated in Fig. 508. It is obvious that more wire is required to make this type of construction than the double cone, and consequently the cost is greater. In the double cone spring construction from 13 to 15 gauge wire is used, and about 32 springs are required for the average-sized back. "Marshall," or nested style springs use 15, 16 and 17 gauge wire and require about 60 springs for the average back.

For custom work, the "Marshall" spring is easily among the best types, as it is made up in convenient lengths and sewn in a continuous cloth sack. These strips can be built up to any desired shape by folding the strips up. In an analysis of the merits of the two types of springs it is easily seen that the cylinder type of spring is the best. For instance, refer to the sketches 509 and 510. In Fig. 509 is shown a piece of cloth laid upon double cone springs which show the areas in shaded lines where there is no support on the spring. The areas shown in Fig. 510 on the cylindrical spring are much less. This gives more support to the passenger's back and at the same time, as there are more springs to carry the load, the wire can be lighter, giving a much easier riding spring.

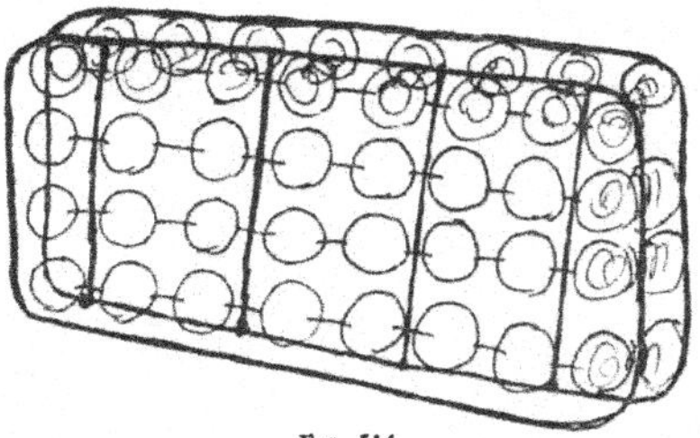

FIG. 514
Verticle brace wires are used in back cushions of the double curve type to prevent them from being bent out of shape.

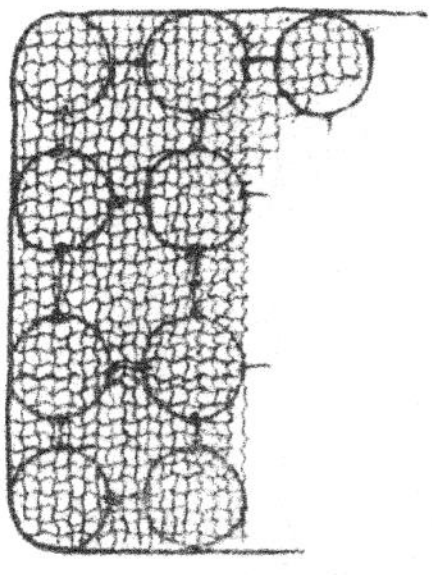

FIG. 515
Fabric woven wire or màttress type tops are being used quite extensively to cover double cone springs.

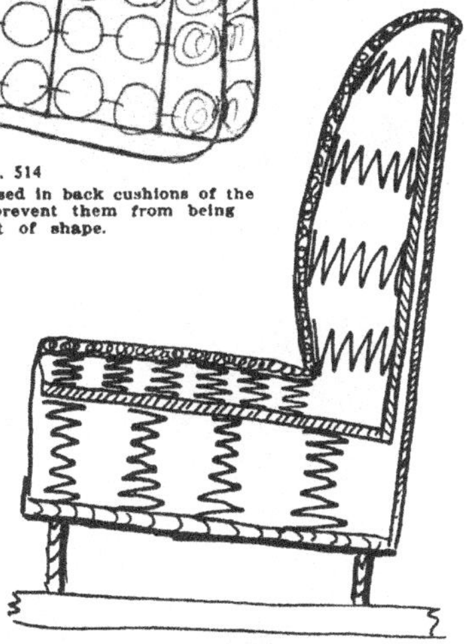

FIG. 516
Combined cushions and back frames have been developed where the back and cushion springs were attached to a frame fitted into groves so it could move with the main load-carrying springs underneath the lower section of the frame.

Back springs, of course, vary in detail according to the manufacturer's ideas. A few of the most interesting variations will be outlined. In the sectional view of the trimmed back shown in Fig. 511 a conventional back spring of either "Marshall" or double cone is set in, and the leather is pulled over it to give the desired shape, and to round over the top loose hair is packed in. In order to economize on the hair several constructions have been developed, two of these are shown in Fig. 512 and Fig. 513.

Formed wires are used on the construction shown in Fig. 512. These are extended to the top of the back and hold up the trimming and hair so that a well formed top is made. In Fig. 513 this same result is obtained in a better way by making the top row of springs form a graceful curve. Back cushion springs of the double curve type are likely to be bent out of shape in service and this

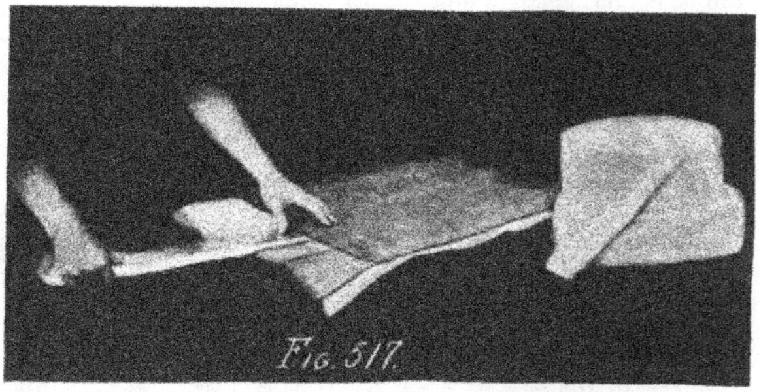

The cotton comes ready for the pipes. This cut shows the cotton and the method of inserting it into the pipes.

Cushion and back pads come in rolls like this and the cotton may be purchased in any width required.

results in a bad appearing back. To overcome this trouble vertical trace wires are used to prevent the springs from falling down as shown in the cross section sketch No. 514. Tie clips are used between each spring to hold laterally.

Fabric woven wire or mattress type tops are being used quite extensively to cover the double cone spring. This is shown in Fig. 515 stretched over the top of the springs and supports the hair and trimming and overcomes the condition shown in Fig. 509, where loose hair is used on the top.

Some combination back and cushion springs have been made where the mattress type covering was stretched over the back and cushion spring, the idea being to make deflections work in unison and overcome chafing the passenger's back. There are two faults with this method, one that the cushion spring, having a greater load than the back spring, tends to pull out of shape, and the other is the difficulty encountered in trimming or covering the springs.

Several styles of combined cushions and back frames have been developed where the back and cushion springs were attached to a frame which was fitted into grooves, so it could move up and down with the main load carrying springs underneath the lower section of the frame. This is shown in Fig. 516. When the seat is wide the springs tend to tilt and cramp when one passenger is riding

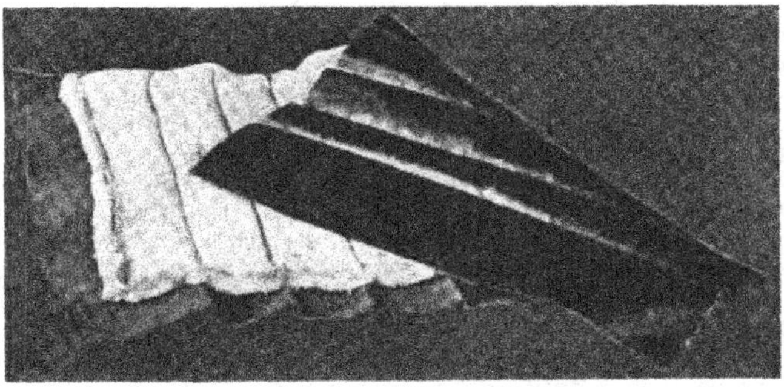

Quilted cotton pad for automobile upholstery.
(The Burton-Dixie Corporation, Chicago.)

on it. It also adds a number of parts to the body, and, as it is hard to move out of the car, it is practically impossible to use the compartment under the seat.

Cushions and back springs are arranged to absorb road shocks but they must be covered with suitable pads to make them comfortable to sit on. Hair and cotton can be used for this purpose. Curled hair has been used in the upholstery trade for a long time, and it is only of recent years that cotton has been fabricated to substitute it. Curled hair is made in various grades and it requires an expert to determine the mixture and the grade. Curled hair must be clean to work, hold together well and be elastic or springy. To satisfy these specifications it must have a large fraction of long horse hairs. Curled hair put on the market is generally composed of horse tail hair. Horse mane hair, cattle tail hair, hog hair, and cattle hair are mixed in accordance with the price it is to be marketed for, and also mixed to give it hard and soft qualities.

The new style of trimming has developed a process of making hair pads. This is loose hair piled up to a uniform height and interlaced or woven to a burlap foundation. These pads can be cut to any shape and will replace the pads or bags of hair which used to be made up to cover back springs. Strips can also be made of interlaced hair the right width and weight to fill out the pipe or plaits of cushion tops or back.

The pipe fillers are inserted into the pipes with appropriate tools and when in place make a uniform filler. This development of pads has simplified trimming operations considerably and resulted in a more uniform output in production work.

Cotton fabricated and mixed with hair is easily worked into pads or sheets, and with normal market conditions can be manufactured and sold at a low cost. It is much used on low priced cars as the cheaper grades of cotton can be worked up to satisfy the conditions required. It is also used on high grade work, in which case the finest cotton is used. On closed body work it makes an ideal pipe filler as it is so soft. Behind the pipes a cotton or hair pad can be used if desired. Cotton can be formed up in strips suitable for the pipe. Fig. 517 shows the cotton used and the method of inserting it into the pipes. Cushion and back pads can be made in any width of shape and come in rolls like that shown in Fig. 518. The pipe fillers are made 3 to 4½ inches wide; weigh 2½ to 6 ounces to the running yard. They can also be made heavier or lighter if desired. Quilted pads have been developed that can be used instead of pipe fillers. The pipe or cotton pocket is dispensed with and the leather sewed up in 4-inch strips leaving off the flaxen or burlap backing. The quilted pad fits right behind the leather and gives the pipe effect. Another method would be to sew the leather right on to the pad as shown in Fig. 519. The cotton pad is sewn between the cotton sheetings so it cannot get away and must give a good evenly fitted pipe effect.

Trimming styles have varied a great deal from year to year. Sometimes the changes were only in detail and at other times were radically different. The prevailing style, and one that seems to meet all the requirements of automobile service, is the French buttonless plait style. Since the adoption of this method of trimming ninety per cent. of the manufacturers have taken it up, and the result is that this style may be termed the standard one for automobiles. The photographs, Figs. 490 and 491, show a typical example of the French plait type of trimming. It will be seen that the plaits are used on the seat backs, sides and the tops of the cushions. The reason for the popularity of the plait trimming is its smooth, clean appearance, and the absence of holes or deep crevices permits it being easily cleaned.

Types of trimming, other than the French plait, that are still being used are shown in the photographs, Figs. 492 and 493. Smooth cushions and backs like that shown in Fig. 492 are very hard to make properly and they have a tendency to show wrinkles as soon as the hair packs down a little.

A "pipe and button" style is shown in Fig. 493. This is similar to the French plait only it has deep depressions where the buttons are located. Buttons are used at the intersections. The reason for dividing the surfaces of the trimming is to make small pockets for the hair so that it will not pack down so readily and become lumpy. On the French plait style the hair is arranged by tufting and the plaits are generally filled with cotton. This works very well and does not become lumpy. As the cushions, seat backs and sides are the principal surfaces to be trimmed, the car's trimming style, therefore, is determined by them.

In order to take up the parts used and the surfaces that are trimmed on a roadster or touring car the following list has been arranged:

BACKS—
- Front Seat Back Trimming;
- Auxiliary Seat Back Trimming;
- Rear Seat Back Trimming.

SPRING OR SPRING CONSTRUCTION—
- Front Seat Back;
- Auxiliary Seat Back;
- Rear Seat Back.

CUSHIONS—
- Front Seat Cushions, Springs or Spring Construction;
- Rear Seat Cushions, Springs or Spring Construction;
- Auxiliary Seat Cushions, Springs or Spring Construction.

FRONT SEAT REAR TRIMMING.

AUXILIARY SEAT TRIMMING.

COWL TRIMMING.

HAND PADS FOR DOORS AND BODY PANELS.

PROPER FORM OF SEAT BACKS.

Seat backs vary a great deal in their contour, and opinion as to just what the back shape should be is very much divided. The writer's opinion is that the

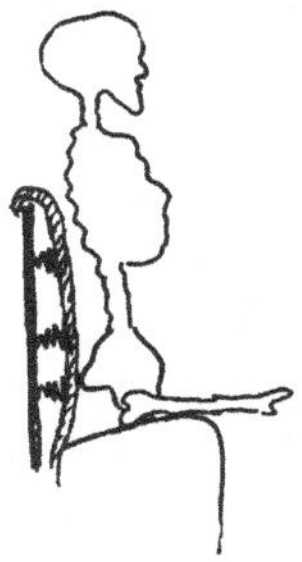

FIG. 520
Correctly shaped seat and wrongly shaped back.

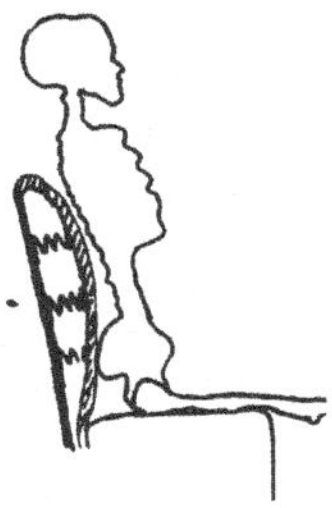

FIG. 521
Correctly shaped back and wrongly shaped seat.

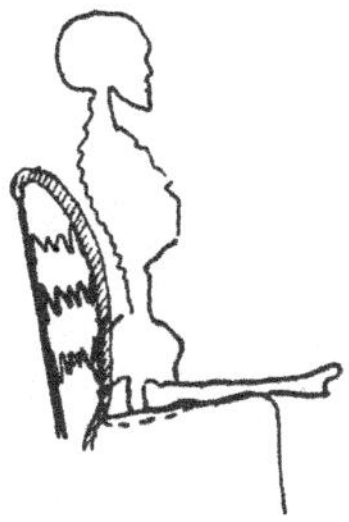

FIG. 522
The right combination—a correct seat and a correct back.

bottom should be slightly full to confrom to the curve in the spine. This applies especially to men. If the passengers are women the back should be almost straight. The back should make a little less than a right angle to the cushion, the cushion being from 1½ to 2½ inches lower at the back than at the front. The width of a front seat which is of normal height should be from 17½ to 18½ inches, while the rear should be from 18 to 19 inches. It is a mistake to make the rear cushion much wider unless it is set very low. The figures are quoted to conform to the average individual.

Where it is possible, the cushions and back trimming should be fitted to the individual. Three seat and back conditions are shown in Figs. 520, 521 and 522.

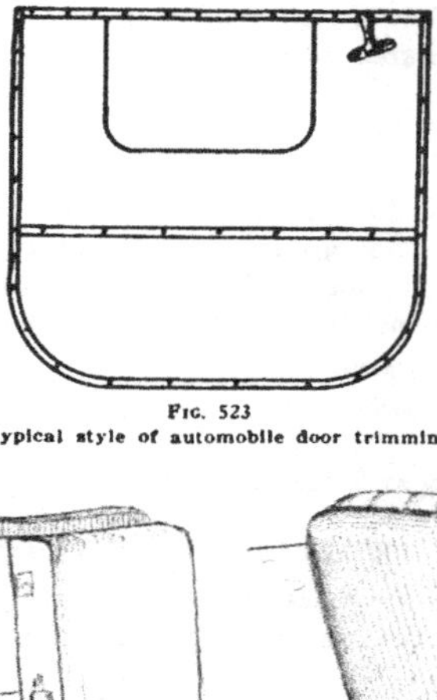

FIG. 523
A typical style of automobile door trimming.

FIG. 524 FIG. 525
Front seat backs are generally provided with a scuff pad, and sometimes they are completely covered with leather or leather substitute. Figs. 524 and 525 show examples of these two methods.

Fig. 520 shows the wrong position for back and correctly shaped seat. In Fig. 521 is shown correct back but wrong seat, while in Fig. 522 is shown a correct seat and back. It will be noted that the mean angle is practically a right angle. The extra support for the spine is obtained by having the row of springs, which comes at a point where the hollow of the back would rest, made a little heavier than the other rows. This will give the correct support when under load and at the same time will not show a lump in the trimming when the seat is not in use.

Seat side and body side trimming is arranged to match the back and cushion trimming, and it is backed by a pad or layer of hair or a substitute. The edges are bound with suitable bindings to make a finish.

Door trimming is generally stretched and pasted over a ragboard foundation, which is tacked to the door and bound with a binding. When pockets are required a cut-out is made in the ragboard and a suitable flap tacked on before the binding is assembled. In Fig. 523 is shown a typical door trim.

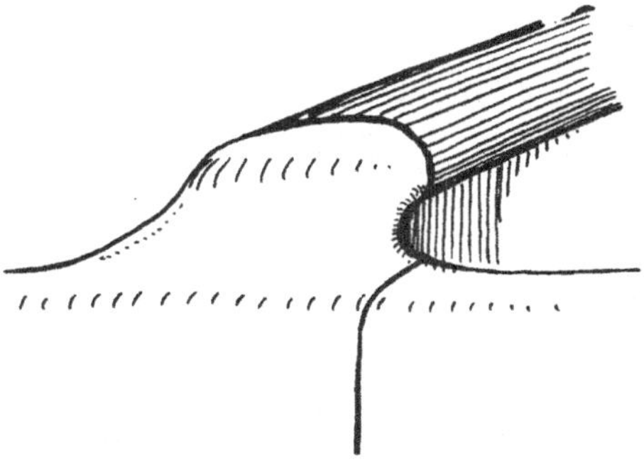

FIG. 526
Top trimming on the front seat that has the double cowl. The leather in this case is stretched tightly over the wood center rail.

Front seat backs are generally provided with a scuff pad and sometimes they are covered completely with leather. Figs. 524 and 525 show typical examples of these two methods. The backs just described are ones used on five-passenger cars. When auxiliary seats are used there is a slightly different condition, as the auxiliary seats generally fold into recesses at the rear of the front seat. Fig. 526 shows the top trimming on the front seat that has the double cowl. The leather in this case is stretched tightly over the wood center rail. The auxiliary seats are generally so small that springs cannot be used, so the back and seat are covered with a hair pad and trimmed to match the rest of the car.

CHAPTER XX

CLOSED BODY DESIGN

At the time of writing this book the closed body development is being governed by two distinct trade conditions. One condition is the special designed bodies that the custom body builder produces for the individual who wants a distinctive car and spares no expense to get the best in appearance and workmanship. The second condition is the large production closed body made by the manufacturers with unlimited amount of resources that make possible low production costs. The first condition, and the one which really laid the foundation of the closed body, produces the luxury car, while the second condition takes from the custom body builder the principles of construction and adapts them to the low priced car, leaving out everything but the parts and trimmings necessary, and then manufacturing the cars in large quantities. Not long ago, to own a closed body was almost the badge of a millionaire, but modern manufacturing has brought the closed body within reach of a very large class of people. Since the automobile is now a necessary means of transportation the closed body supplies the real demand for a car when all round serviceability is concerned.

The closed body is really the automobile industry's biggest problem because the car the people need the most is the closed car, and as soon as the engineers and manufacturers can provide enough cars at an acceptable price the open car will be superseded and placed in the special or luxury car class. In Figs. 527 and 528 the photographs are shown of two cabriolets, typical examples of custom made body work. They are designed to be driven by a chauffeur and have the rear adaptable so it can be opened up in good weather. In Fig. 529 is shown a special sedan. The landaulet, limousine, brougham, berline and similar constructions come under the luxury or "special" closed body types. The closed bodies which are now being manufactured in large quantities could be practically called two types, the coupé and sedan, or 4-passenger closed body and seven-passenger closed body. Most coupés have a seating arrangement whereby four passengers can be carried, and the sedan will carry six on the small chassis and 7 or 8 on the large chassis.

The automobile of the future will be in two classes—the close-coupled four-passenger closed car, and the seven-passenger closed car. All the other cars will be in the "special" class. The photograph 530 shows a typical four-passenger car on a six-cylinder chassis and photograph Fig. 531 a seven-passenger car. A good example of the existing demand for automobiles with closed bodies is shown

Cabriolet, built in a custom shop. The rear can be opened up in good weather

Custom built cabriolet. Both cars on this page are designed to be driven by a chauffeur

FIG. 529
Design for a sedan, which won first prize at the exhibition of the Automobile Body Builders' Association in New York City, January, 1922. Designed by Philip Brostrom, Long Island City, N. Y.

Typical four-passenger car, made in quantities. This body is on a six-cylinder chassis, 124-inch wheelbase.

in the 1922 line of the Buick closed cars shown in Figs. 530 to 535. The coupé and sedan shown in Figs. 530 and 531, as previously described, are four-passenger and seven-passenger models. Fig. 532 is a three-passenger coupé, and Fig. 533 is a five-passenger sedan. All of these models are in six-cylinder chassis, the three-passenger coupé and five-passenger sedan being on 116-inch wheelbase chassis, and the four-passenger coupé and seven-passenger sedan on a 124-inch chassis. The coupé, Fig. 534, and sedan, 535, are on a 109-inch wheelbase four-cylinder chassis.

In Fig. 536 is shown a picture of an unfinished four-passenger coupé. This gives a good idea of the construction. The panels are all steel excepting the post covering, where aluminum is used because it is easier to handle. The roof is made of composition supported by slats. A seven-passenger sedan is shown in Fig. 537. This is designed to permit removing the center and rear door posts so that a clear vision can be obtained when desired. This is a four-door body and it entails the use of many parts to get the clear vision. In Fig. 538 is shown a five-passenger sedan with only two doors. This has detachable posts and works out to better advantage than the four doors. In Fig. 539 all the posts are removed so it can be seen how open the job is made under this condition. At one time it was thought it was desirable to have this clear vision construction, but it invited so many complications in its construction that it was expensive. It also weakened the body to have the removable posts. The result is that there are very few bodies made now with these detachable features.

NOMENCLATURE OF CLOSED BODIES

As the closed car had developed from the horse-drawn vehicle it is quite natural that many of the names for closed bodies are derived from their prototypes used before the automobile was a reality. While on the subject of names it is worth mentioning in order to avoid discussion, that the modern closed bodies have suffered severely in being christened. A number of names have been applied and some accepted, which really were wrong in the light of previous practice, but as they have been generally accepted the easiest thing is to use them. The Society of Automotive Engineers has suggested the standardization of the following names for closed cars:

Coupelet—Seats two or three. It has a folding top and full height doors with disappearing panels of glass.

Coupé—An inside-operated enclosed car, seating two or three; a fourth seat facing backward is sometimes added.

Convertible Coupé—A roadster provided with a detachable coupé top.

Convertible Touring Car—A touring car with folding top and disappearing or removable glass sides. ("Phaeton" has been suggested recently as a better name than "touring car.")

Sedan—A closed car seating four or more, all in one compartment.

A seven-passenger car manufactured on a large scale. Six-cylinder chassis, 124-inch wheelbase

Three-passenger coupé, on six-cylinder chassis. An excellent example of a large production closed car. 116-inch wheelbase.

Five-passenger sedan on six-cylinder chassis, 116-inch wheelbase

Coupé body on four-cylinder chassis, 109-inch wheelbase.

Convertible sedan—A salon touring car provided with a detachable sedan top.

Open sedan—A sedan so constructed that the sides can be removed or stowed so as to leave the space entirely clear from the glass front to the back.

Limousine—A closed car seating three to five inside, with driver's seat outside covered with a roof.

Open limousine—A touring car with permanent standing top and disappearing or removable glass sides.

Berline—A limousine having the driver's seat entirely enclosed.

Brougham—A limousine with no roof over the driver's seat.

Landaulet—A closed car with folding top, seats for three or more inside and driver's seat outside.

In order to make sure that there is no misunderstanding, the illustrations from 540 to 554 show the constructions which compare with the Society of Automotive Engineers' names.

The various body types will be examined in the order that the names were arranged. First of all comes the coupelet. In Fig. 540 is shown a good example of the modern design of this type, and in Fig. 541 is shown an earlier type. While the general construction is the same the latter type has been worked out to suit the trend of design. The seating arrangement is the same, the main difference being in the back of the top. This is made to have a flat roof on one and a rounded back and roof on the other. The rounded back is covered with leather while the flat top has a duck covering. The rear door pillar is made detachable, and when it is removed, the top let down and the door glasses lowered, the body is as open as a roadster.

The coupé shown in Fig. 542 has a permanent roof with the front and rear side windows made to operate. The rear door posts can be made in two ways, removable and stationary, the latter method being better and more modern.

A convertible coupé is shown in Fig. 543 and is a construction which is so designed that the upper section or top can be detached at the belt line, allowing the car to be used as an ordinary roadster. In order to have a satisfactory top the lower section of the body must be specially constructed so that it will fit on and make a good joint. At its best it does not compare with a regular closed body, but is an improvement over the "winter top."

The convertible touring car as shown in Fig. 544 is practically obsolete, and was the original Springfield type of body. The sedan classed as an open sedan was marketed as its development. The sedan body shown in Fig. 545 is designed to seat from four to five and sometimes six passengers, all of them being in one compartment. It is a permanently closed body with either two, three or four doors. The earliest designs had a divided front seat, which allowed the eliminating of two doors, having one door at the front left and one at the rear left. By having but two doors the body construction was made much lighter and stronger. The modern type of sedan body is made with four doors and a solid front seat.

The reason for this is that it will allow the passengers getting out much easier from either side of the car. A divided front seat would not leave very much room for passage between them, therefore there was a certain amount of inconvenience attached to a two-door body. Another feature in favor of a solid front seat and four doors is that in case of an emergency three passengers can be accommodated in the front seat. The windows in each door can be lowered, and also the windows on each rear side of the rear seat.

An open sedan, which is shown in Figs. 538, 539 and 546, is sometimes termed a Springfield type body, but there is a point that could be used for a distinction between them. The difference is in the manner the side windows are closed. In the sedan the pillars are removable between the doors, and the windows drop down in the side of the frame. While the Springfield type construction has

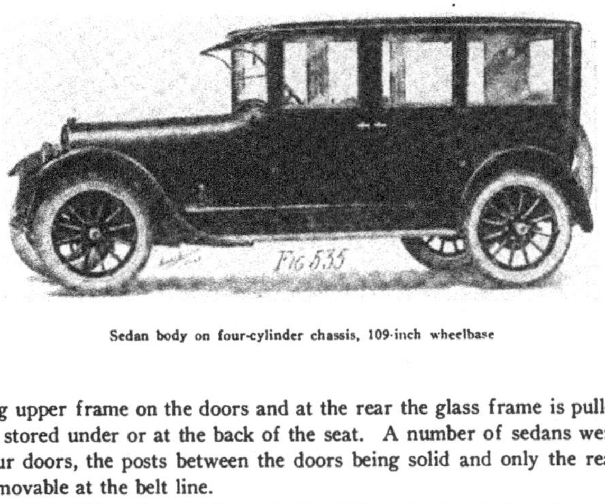

Sedan body on four-cylinder chassis, 109-inch wheelbase

a folding upper frame on the doors and at the rear the glass frame is pulled right out and stored under or at the back of the seat. A number of sedans were made with four doors, the posts between the doors being solid and only the rear posts were removable at the belt line.

The convertible sedan, as shown in Fig. 547, is designed along the same lines as the convertible coupé previously described.

The limousine shown in Fig. 548 is the "exclusive car" and it is made for city use, to be driven by a chauffeur. The passenger compartment is made to seat three on the rear seat and sometimes has two folding auxiliary seats. The top is extended over the driver's seat and a short door, like a touring car door, is fitted. The windows in the doors and side can be operated at will. The open limousine, sometimes called the touring limousine, as shown in Figs. 549 and 550, is built along the same lines as a sedan, with four doors and solid driver's seat, and differs mainly by having a glass behind the front seat that can be raised or lowered, when closed dividing the car into two complete compartments.

Four-passenger coupe in course of construction. Panels are all steel except the post covering, where aluminum is used. Composition roof, supported by slats

Seven-passenger sedan in course of construction. This is a four-door body and requires the use of many parts to obtain clear vision.

Five-passenger sedan with only two doors. This has detachable posts and works out to better advantage than the four doors.

Same job as above with all the posts removed so it can be seen how "open" the body can be made when desired.

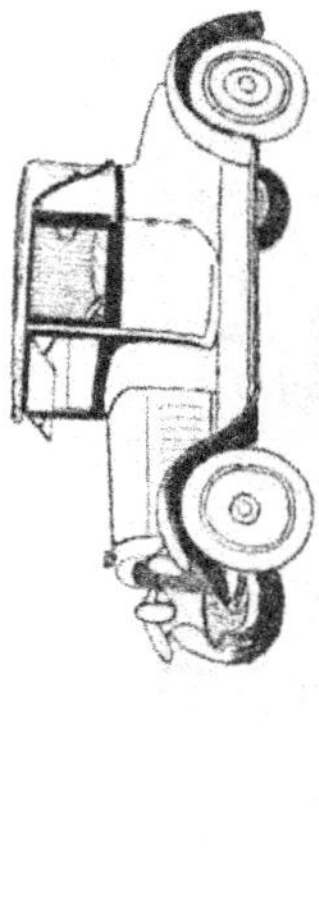

FIG. 540
Modern coupelet

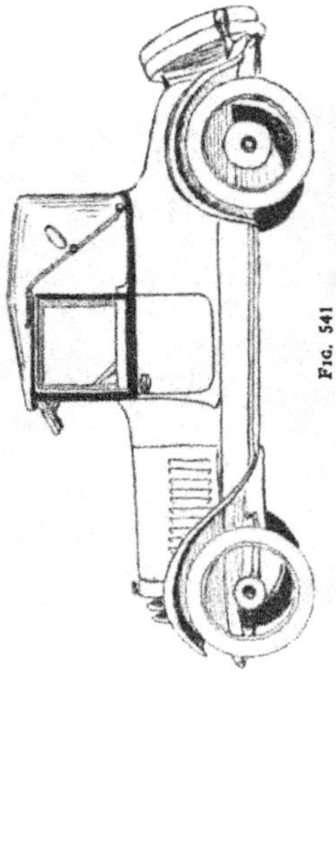

FIG. 541
An earlier type of coupelet

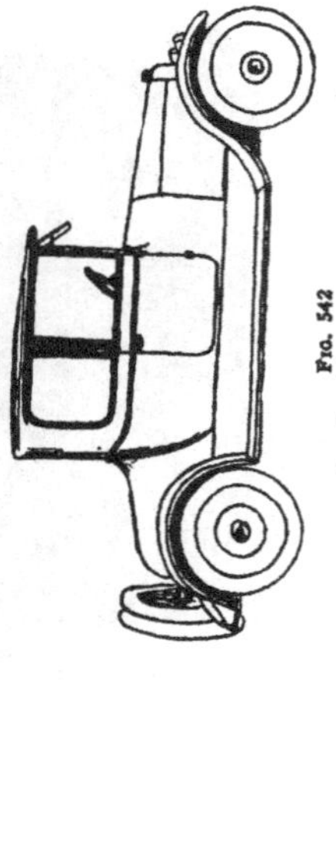

FIG. 542
Coupé, with permanent top

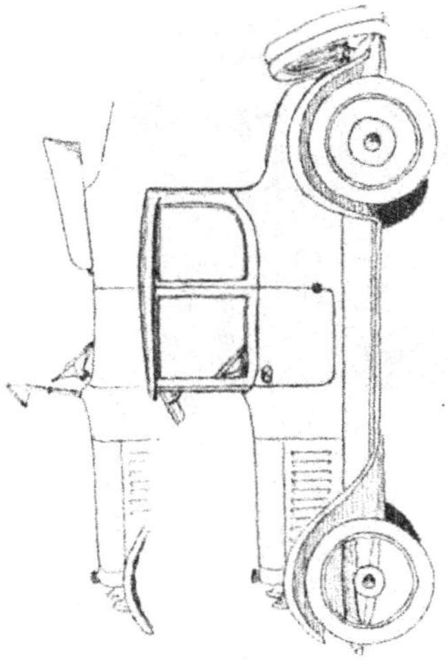

FIG. 543
Convertible coupé, with upper section detachable at the belt line

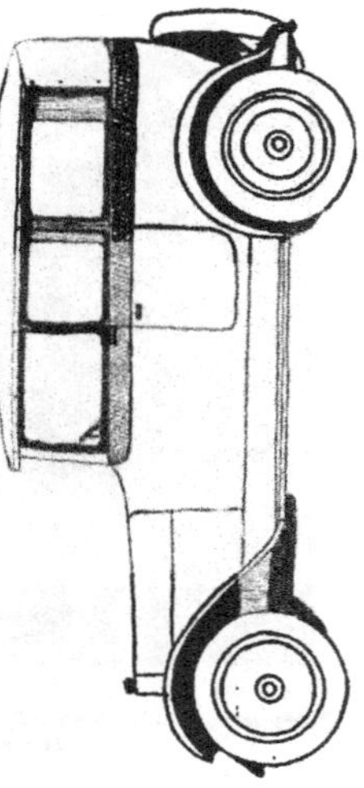

FIG. 544
"Convertible touring car", a type now virtually obsolete

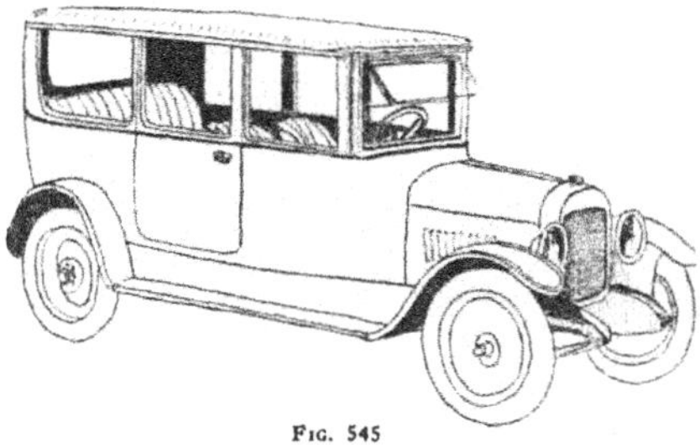

FIG. 545
The sedan body, which accommodates all the passengers in one compartment

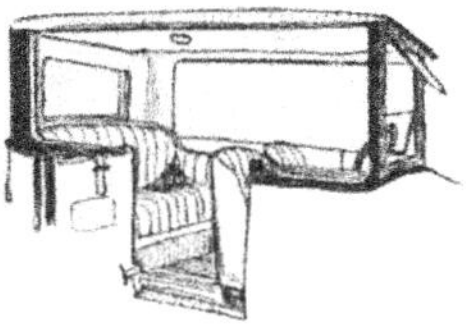

FIG. 546
So-called "Springfield" type sedan.

FIG. 547
Convertible sedan, with upper section detachable at the belt line

FIG. 548
Limousine body, designed for city use and to be driven by a chauffeur

FIG. 549
A "touring" limousine, having a glass behind front seat that can be raised or lowered

FIG. 550
In the "touring" limousine, when the glass is closed, the car is divided into two complete compartments

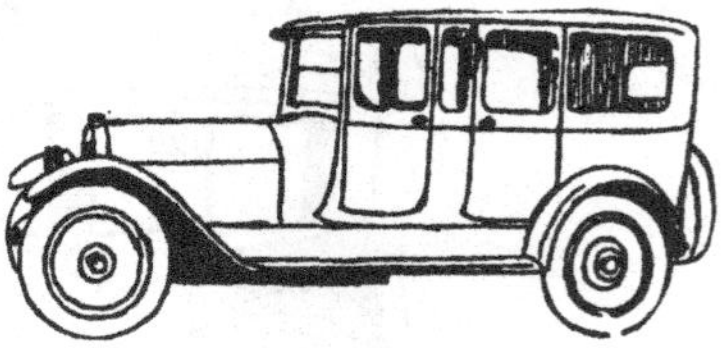

FIG. 551
The berline has the chauffeur's compartment permanently divided from the rear or passenger compartment

FIG. 552
The brougham is generally built with a smaller rear compartment than the limousine, and is often called a "town car."

There is but a small detail in the variation between a touring limousine and a berline. See Fig. 551. The latter has the driver's compartment permanently divided off from the rear, thus giving limousine accommodations at the rear and full protection to the driver at the front.

FIG. 553
The landaulet is a closed car along the same lines as a limousine, but the rear part of body can be folded back. Sketch shows rear opened up

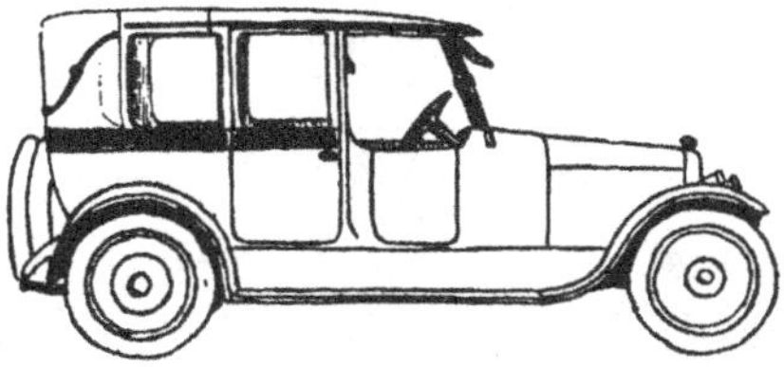

FIG. 554
This sketch represents the same body as the one shown in Fig. 553 above, but with the rear closed, fully protecting the passengers in the worst weather

The brougham is generally built with a smaller rear compartment than the limousine and is sometimes called a town car. It is a type very popular at present and much used by women. A typical design is shown in Fig. 552.

The landaulet is a closed car along the same design as a limousine, with the door and side windows operating. The back end of the body can be folded back, giving an open car effect at the rear. This type of body is adapted for city use and for driving in the park. In Fig. 553 is shown the landaulet with a back opened, and in Fig. 554 it is shown closed.

CHAPTER XXI

PRINCIPLES OF TRUCK BODY DESIGN

The development of the gasoline engine, like the invention of the steam engine, has revolutionized the world, through its assistance in annihilating or reducing distance.

A city which is but a short distance from another is quickly reached by motor transportation. Commercial transportation had to rely upon the steam railroads for its handling and in comparatively short distances several days would elapse in getting freight cars loaded, started, located and unloaded. With a motor truck the material can be hauled direct from door to door at a considerable saving of time.

Fig. 555.
Example of open express body with top over seat

Then again there is the town or village a few miles from the railroad, which had to rely upon hauling by teams to and from the depot, a motor truck can do this trucking in a fraction of the time a team would take.

A short time ago a strong line divided the truck from the passenger car, but now buses are made from trucks and so designed as to carry twenty or more

FIG. 556.
Same type of express body as shown in Fig. 555, but with canopy top

people in them. Motor truck body styles are not much different from those used on horse-drawn vehicles, but unlike their brothers the passenger cars are not subjected to the modification of season style changes. They are, however, going through a process of evolution as new ideas are worked out from time to time which make the truck better fitted to the purposes for which it is intended, or improves the construction so it will give better service.

The truck body is a big problem for the engineer, as each different trade or business feels that it has to have a body equipment that suits their purpose. This means that all trucks above the light class, demand more or less special equipment.

FIG. 557.
Typical light delivery or package body, with flare boards and curtains at side and rear

Some of the different businesses that utilize trucks are dry-goods, lumber, oil, furniture, coal, ice, building trades, packers, laundries, undertakers, farmers, milk, flour, florists, etc. The army uses a number of different types of trucks, and every large city has to have an equipment of trucks adaptable for miscellaneous purposes around the city.

Fig. 558.
Four-post light delivery body with flare boards and curtains side and rear

TRUCKS CLASSIFIED INTO THREE MAIN GROUPS

In order to conveniently study the different types of trucks, three groups will be made. The first group covers the express type body; the second group the platform and its combinations, and the third, special body construction. The express body is the most adaptable to any commercial type body, and it is made in a number of sizes and varies greatly in the details of its construction. There are two divisions; first, the open express; and second, the closed express. In Figs. 555 and 556 are shown examples of express cars. The first is the open body, and the second the same body with a canopy top fitted to it.

THE EXPRESS OR PARCEL BODY

The first class of commercial car to consider is the small four-cylinder car, that has a light express body fitted to it, and is used for grocery stores and similar business purposes where the loading is light. This class of body or car would well be called a package wagon. In Figs. 557 and 558 are shown typical four-cylinder light delivery or package delivery cars. It is of very simple construction and can be made very light so that it can be used on a small chassis. The driver's seat is attached to the front end, which also has the toe board risers and floor boards attached, making it an easy body to assemble on a chassis.

The construction consists of sills, floor boards, side boards, flare boards, and end or tail gate. A partition in the front end provides a place for the gas tank and tools, and the cover is formed by the seat frame.

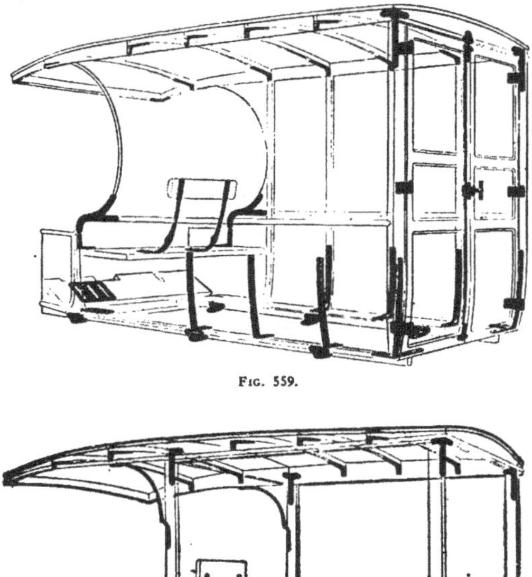

FIG. 559.

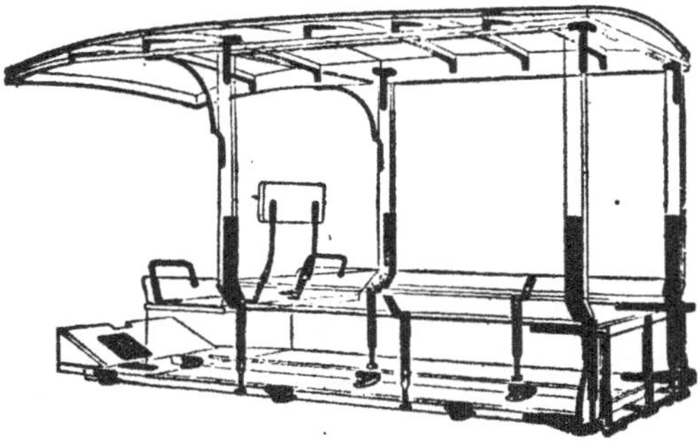

FIG. 560.

Figs. 559 and 560 show skeleton views of a panel express body and an open express body with canopy top. Both drawings show the different locations which require brace irons.
(Courtesy Eberhard Mfg. Co., Cleveland, Ohio)

The hardware, which is always a very important item on a truck body, consists of floor board plates of strip steel; flare board and tail gate binding of strip steel; flare board and side board braces of malleable iron; tail gate hinges of malleable iron; and tail gate chain with malleable iron hooks. Fittings for trucks

FIG. 561.

In this car a passenger chassis is used; and front end, consisting of cowl, windshield and front door are worked into the commercial body construction

BODY DATA SHEET

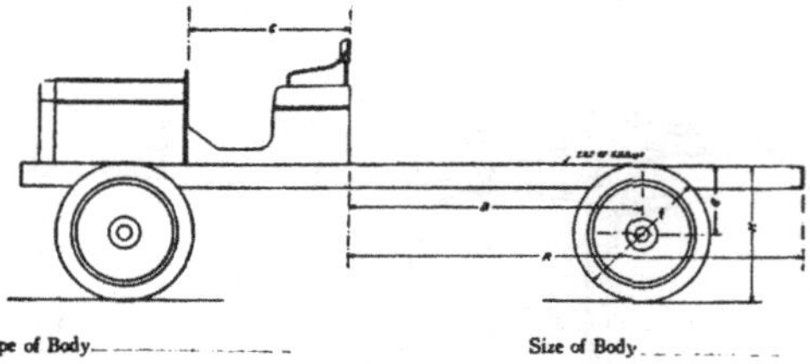

Type of Body ________ Size of Body ________

A ________ F ________

B ________ G unloaded ________

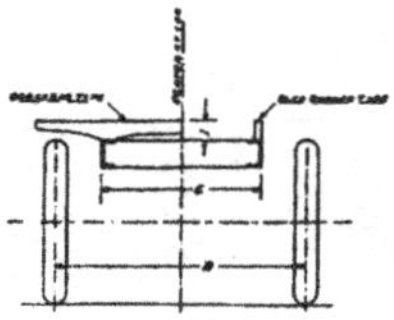

C ________ H unloaded ________

D ________

E ________ I ________

Type of subsills—Sled Runner, Cross Bar.

Customer's Name ________ Customer's Order No. ________

Make of Truck ________ Capacity ________

Our Order No. ________ Our Cost No. ________ Date ________

The above data should be given to enable us to get out and attach subsills

FIG. 562.

Reproduction of a body data sheet used by a well-known truck body manufacturer are manufactured by several large concerns who have a complete line of parts adaptable to most designs.

A canopy top is almost indispensable for a truck that will have constant service. This top is supported by four or more posts, a construction which is satisfactory for short bodies. The top consists of wood frame with cross ribs and slats running lengthwise. The slats should be close enough together to support the top covering without the slats showing through. For a covering a ten-

FIG. 563.
Open express body with chain tail gate

FIG. 564.
Express body with canopy top. This is a four-post body, but additional posts may be added if desired

ounce black duck material can be used. For side curtains a lighter duck can be employed. An eight-ounce fabric will be found to give good service. In fitting up a top it is very essential that well-designed brace irons be used in the corners where the posts attach. Another point to be watched is the attachment of the posts to the body. These should be fastened very securely and can be clamped with brackets.

Unless the body maker or car manufacturer intends to produce large quantities of truck bodies it would not pay him to make up patterns and equipment to

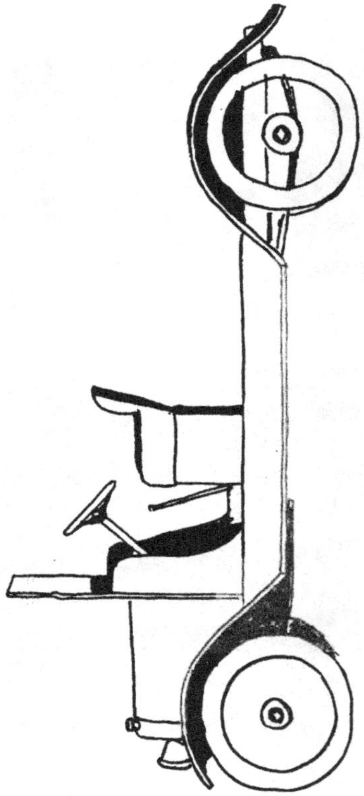

Fig. 565.
Chassis as furnished to body builders with seat and tool box

manufacture miscellaneous truck body irons, when they can be obtained from stock from the companies which specialize in making the parts. In Figs. 559 and 560 are shown skeleton views of an open express body with canopy top and a panel express body. These drawings show the different locations which require brace irons. All the irons and parts shown can be obtained from the stock of the parts manufacturer.

The open express body shown in Fig. 560 is made so that the canopy top need not be used unless required. It requires the following parts:

Size 10-inch side panel; 5-inch flare board
6 pieces body brace and flare board irons
4 pieces flare board and post irons
3 pieces end gate hinges
1 pair end gate springs
2 pieces end gate spring guards
2 pieces body brace, rear end
1 pair seat brace and handles
2 pieces corner irons, top of seat
1 pair jointed lazy back irons
6 pieces body and chassis connections
1 set footboard plates
2 pair top rail corner irons
10 pieces roof rail corner irons
2 pieces top rail corner plates, front
2 pieces top hood braces
1 pair sill braces

It will be noted that the canopy top irons on the posts are made so that they can clamp right on the flare board of the body. The panel body shown in Fig. 559 is a complete unit and it is much better to make this with solid sides than try to make a top which is detachable from the base. The following is a list of irons, which are shown:

1 set Acme door lock
4 pieces door hinges, upper and middle
2 pieces door hinges, lower
1 pair rear post, inside body braces
6 pieces side post, inside body irons
8 pieces roof rail corner irons
1 pair front arch stay braces
1 pair rear post, top rail corner irons
2 pieces top rail corner plates, front
2 pieces seat and body corner irons
1 pair joint lazy back irons
1 set foot board plates
6 pieces body and chassis connections

FIG. 566.
A speed wagon type, with removable top and posts, easily convertible into an open body

The open express type of body is very adaptable and even in small sizes can be used for a number of purposes. A common addition to this body is to add wire mesh or expanded steel to the sides so they will be completely enclosed. This allows packages and boxes to be piled the full height of the inside of the truck without danger of their falling out. This type of body is shown in Fig. 560. It will be noted that side curtains are provided to protect the load in rain. It is common practice for the car manufacturer to use the touring car chassis with a few changes for light package wagons. In some cases the front end, consisting of

FIG. 567.
Another speed wagon, with stationary posts and canopy top

cowl, windshield and front door, are worked into the body construction as shown in Fig. 561. There is one objection to this practice and that is the touring car cowl and seat arrangement take up more room than seat construction made like that shown in Fig. 557. It is generally necessary to obtain as much room as possible in the truck body, so it becomes important to cut the seat and leg room down to the minimum.

FIG. 568.

FIG. 569.

Figs. 568 and 569—Full panelled bodies are used on speed wagons. Fig. 568 has driver's seat protected by side curtains, while Fig. 569 shows the driver's seat fully enclosed

When a truck body is to be ordered for a chassis it is necessary to furnish the manufacturer with the general chassis dimensions. For reference and general information the chart shown in Fig. 562 covers the information required. This chart is a copy of one used by a well known truck body manufacturer.

Recent development in trucks seems to point to the medium weight or what could be called "light trucks" becoming the most popular. One of the reasons for this is that the development of pneumatic tires has made it possible to carry heavy loads, which had to be carried by solid tires in the past. The use of pneumatic tires for light trucks has demonstrated a much greater increase in all around efficiency. The loads are carried with less shock to the truck and much higher speeds are possible.

The cord tire which is now being used so much gives good mileage even on the heavier trucks when used. The increased speed possible on trucks with pneumatic tires has made the medium weight truck one of the most used types, if not the most used type on the market. The speed wagon chassis is generally

Fig. 570.
Open express body, with enclosed cab for the driver

of ample proportions so bodies with good loading space can be fitted to them. It is also possible to build a large variety of different styles of bodies for them.

Like the smaller trucks, the express type body is the most used because of its great adaptability and also because the manufacturer can make this body in quantities and satisfy a large percentage of the demands from the trade. The general run is similar to the smaller bodies, but it is heavier and generally has six posts. In Figs. 563 and 564 are shown express bodies with and without canopy tops. This is only a four-post body, but extra posts can be easily added if required. One feature in which the heavy and light bodies differ is that the heavier body is generally made as a unit without the driver's seat, tool-box or floor board. The reason for this is that the speed wagon chassis should have a driver's seat in it complete as first made so that it can be sold that way if necessary. This arrangement will permit any type of body being fitted to the chassis without disturbing the front end at all. The chassis unit with seat will look like Fig. 565.

In Figs. 566 and 567 are shown two of the designs of speed wagons. Body constructions permit the top and posts being removed as the posts are set into the sockets and are held by a bolt which can be easily loosened and the top pulled off. Side racks or extensions for grain and other purposes can be readily fitted by utilizing these sockets.

Full paneled bodies of all kinds are used on the speed wagons. Two types are shown in Figs. 568 and 569. The design shown in Fig. 568 has open sides in the front. In Fig. 570 is shown an open express body with an enclosed cab for the driver. This is a very practical arrangement, especially for the States which experience bad winters.

Trucking is a very old art and the horse drawn vehicle used for hauling all kinds of material and merchandise can give us plenty of pointers in design. There is, however, one thing to remember, that is the difference between the speed of a horse drawn vehicle and a motor propelled one; consequently, there are road shocks and vibrations which the motor truck encounters that never occurred in a horse drawn vehicle. The aim of every truck builder is to supply a truck which will carry as heavy a load as possible at the lowest cost.

FIG. 571.
Truck with driver's seat and cab, with a typical platform mounted in the rear

Starting with the truck chassis itself, it must be as light as it can be made, and yet be capable of withstanding the service and load it is designed to carry. In trying to develop a truck which will stand up, some designers have not spared expense or material, and in some cases the truck has a good weight to move in its own component parts before any load is applied. In making a body it must be sturdy and strong but it must be light as possible, so the problem is not how much can be put on the body but how little and still obtain satisfactory results.

The second group of truck bodies covers the platform and stake bodies. Platform truck bodies are used in the heavier types of trucks although the speed wagon and rapid transit trucks can use the smaller platform to advantage in some businesses. There are some truck body manufacturers who make a platform body so that it will serve as a foundation for a large number of different body styles. This is a very practical construction, both from the sales and from the manufacturing standpoints. It enables a dealer to make prompt deliveries without having to carry a large stock of bodies, thereby permitting them to handle the bodies for a less cost to the consumer.

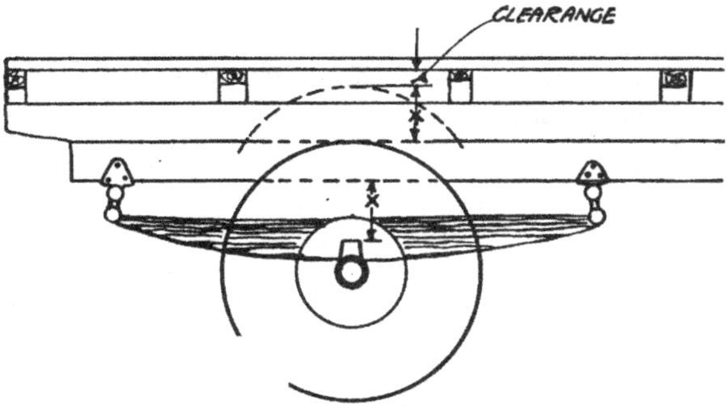

Fig. 572.
Clearance provided to raise the body 5 or 6 inches higher than the ordinary express body

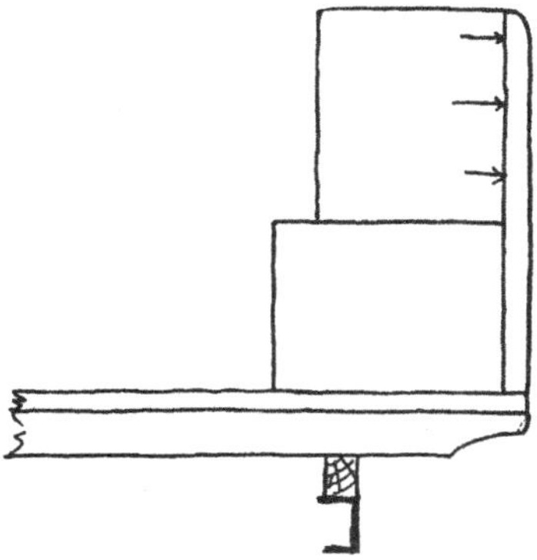

Fig. 573.
Side strains often exert a leverage by being imposed against the stakes, as shown here

The platform body is really a combination of platform and suitable stakes or sides fitted to it. The platform is really the basis for the large varieties of fittings which when combined with the platform define the name of the body type. In Fig. 571 is shown a truck with the driver's seat and cab with a typical

Fig. 574.
New York Stake Body

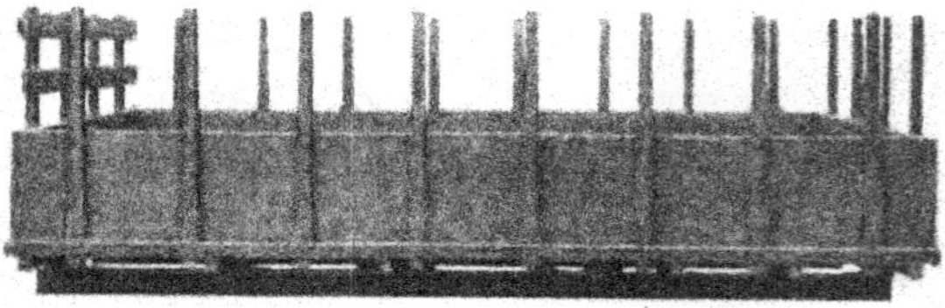

Fig. 575.
Cleveland Stake Body

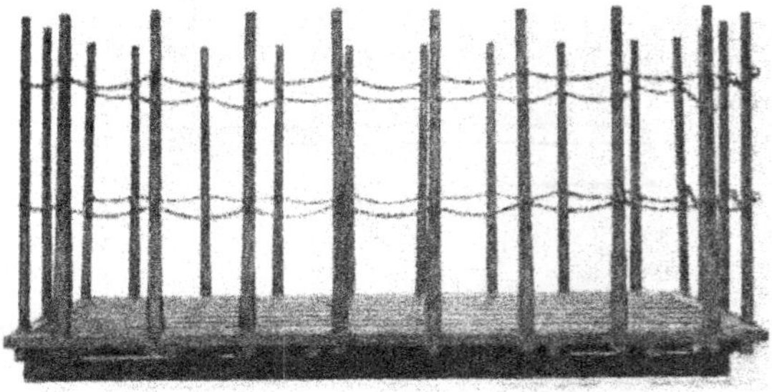

Fig. 576.
Cincinnati Stake Body

FIG. 577.
Boston Stake Body

platform mounted at the rear. This construction differs from the express type principally in the greater width that can be obtained with it. This width is gained by extending the platform over the wheels to a width equal to the width of the general standard for fenders, which is 66 inches. Of course, the width can vary to suit the body requirements and can be made as wide as required, but for general use they should not exceed 70 inches or else they will be difficult to handle in narrow streets. In permitting the body to extend over the wheels, allowance must be made to have it high enough to clear the wheels when the spring compresses. Extra space must be figured on for chain and mud hook

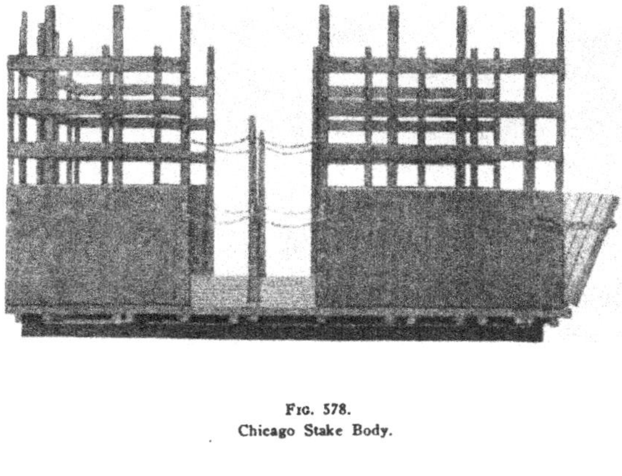

FIG. 578.
Chicago Stake Body.

clearance. By providing this clearance, as shown in Fig. 572, the platform or bottom of the body is raised 5 or 6 inches higher than the ordinary express body.

When the body shown in Fig. 571 is examined it will be seen that the platform is supported by cross sills which rests on sills which fit along the frame.

FIG. 579.
A good design that can be accommodated to a wide range of industrial uses

FIG. 580.
The main difference between this and Fig. 579 is in the wide base board and in the absence of connecting locks or hooks

These are called sub-sills, and they are used to raise the body so it will have the proper wheel clearance. The platform consists of an outer frame with substantial floor boards. The rear end sill has to be made especially rugged and be well braced, because in loading and unloading it has to stand a lot of shocks. On both sides and across the rear end pockets are provided to carry the stakes.

Fig. 582.
Malleable Iron Hook Type Fastener

Fig. 583.
Platform which serves as the foundation for bodies such as are shown in Figs. 584 and 585

Fig. 584.
Platform with sides added

Fig. 585.
Platform with sides and canopy top added

Fig. 586.
Platform showing top view

Fig. 587.
Express Body Sides

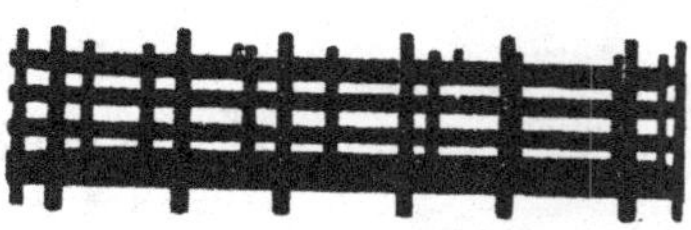

Fig. 588.
Stake Sides

There are several methods used to make stake pockets. One is to make them of malleable iron and rivet them on the inside of the reinforcement that goes round the outside of the body. Another way is to have malleable iron pockets bolted or riveted on the outside of the frame re-enforcement. Some bodies have holes mortised in the wood frame.

The success of the body depends on the method used for the framing and attaching the stake pockets, because the frame supports all desired loading, and

Fig. 589.
Stakes and Gate Ends for Cattle Transportation

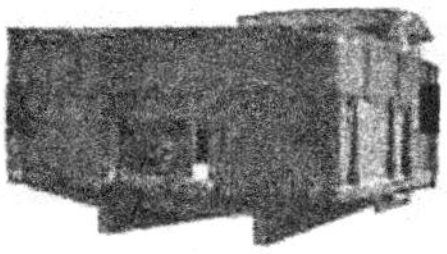

Fig. 590
Sides and Ends for Carrying Grain

the stakes have to support all the side strains that the load imposes. These side strains often exert a leverage by being imposed against the stakes as shown in Fig. 573.

The arrangements of the stakes have a number of names given them in accordance with their popularity in different sections or towns in the country. The New York, Cleveland, Cincinnati, Boston and Chicago are shown in Figs. 574, 575, 576, 577 and 578 respectively. The New York type shown in Fig. 574 has two, three or four sets of stakes with a board across the front two stakes. The Cleveland, shown in Fig. 575 has a wide side board running along the bottom of the stakes. In Fig. 576 is shown the Cincinnati type. This has single stakes and sometimes has a name board attached to the two front stakes.

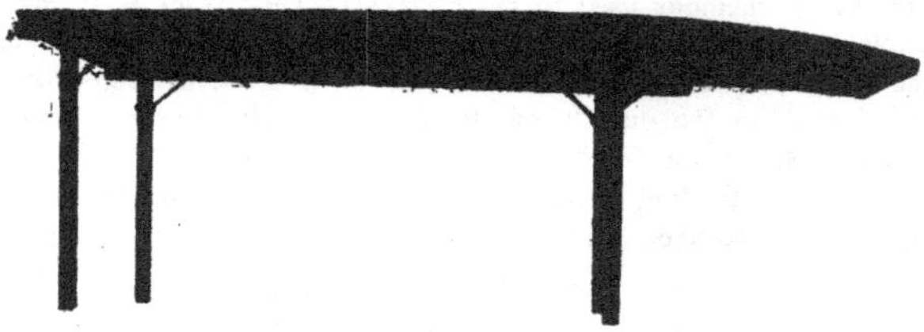

FIG. 591.
A Detachable Canopy Top

The Boston type shown in Fig. 577 has double connected stakes with a space left between them. The front two stakes having a name board attached to them. In Fig. 578 is shown the Chicago type; this has front and rear combination stakes with high extension and solid board at the bottom. A space is left between the combination stakes where a single stake is placed, the front center single and rear stakes being connected with a chain. In Fig. 579 is shown a good design of stake body which can be accommodated to a wide range of trades. The mounted body shown in Fig. 580 is very similar to that shown in Fig. 579, the main difference being in the wide base boards and absence of connecting locks or hooks.

It is good practice to use a hook or lock to prevent the plates from jumping out when traveling over rough roads. A malleable iron standard lock is often

FIG. 592.
Express Type Sides fitted to a heavy truck for hauling sand and gravel

used and in Fig. 582 is shown a hook type fastener. Some makers fit the stakes tight so they can not work out and do not put the hooks on them.

Some of the possibilities of a stake body are shown in Figs. 583, 584 and 585, which show an express body built up in the platform by mounting sides and then a canopy top. In Figs. 586 to 591 are shown a complete combination suitable for general use, but more especially for the farmer. Fig. 586 is the platform. Fig. 587 express body sides, Fig. 588 stake sides, Fig. 589 stakes and gate ends for cattle transportation, Fig. 590 sides and ends for carrying grain and Fig. 591 a canopy top. Photograph 592 shows express type sides fitted to a heavy truck, making it adaptable for heavy work like hauling dirt and gravel.

CHAPTER XXIII

SPECIAL TRUCK BODIES

In previous chapters the common or general type of truck body has been taken up and grouped as well as possible, and all the body types not described have been left for this article. This group is the third section which was arranged for consideration. There are body designs without number that have developed to accommodate the different lines of business. In fact most manufacturers of medium heavy trucks market their chassis without bodies and have bodies made to suit the purchasers' requirements. The large truck manufacturer keeps an organization especially to accommodate the purchasers' requirements in body construction and equipment. For example, fire engines, oil tanks, water wagons, water sprayers, road cleaners, and dumping equipment, telephone repair equipment, light wire equipment, wrecking and hoisting equipment. All the equipments mentioned require mechanical engineering and cannot be readily classed with body design, but they have to be considered in connection with motor trucks.

The construction of special bodies will be governed by their design. Most special bodies will be built with the sub sill bases and platform construction as a foundation. Panel and closed truck bodies will be narrow and governed by the same conditions as the express body which is narrow enough to provide wheel clearance. Transportation of persons and material is the only reason for trucks being made, and for carrying a large number of people the truck has been used to good advantage.

There is one direct connecting link between the motor truck and the body engineer, and that is the development of sight-seeing buses, omnibuses, jitney buses and touring vans. All of the above bodies are being extensively used and mostly on ordinary truck chassis, and in the writer's opinion there will be a big development along the lines of equipment of vehicles for transportation used as supplements to the railroads.

The modern sight-seeing bus has a long body with side doors and front end similar to touring car, only it is much longer. A long fabric top is fitted, also side curtains, making it a complete car, only on a large scale, similar to that shown in Fig. 593. This type of bus is used in those sections of the country which offer picturesque or novel scenery that will interest sightseers and take them out of the city. For city use a more simple construction is used, a long body with seats crosswise on it, without doors or top. These are patterned after the buses used

FIG. 593
A modern forty-two passenger sight-seeing body.
(FitzGibbon & Crisp, Inc., Trenton, N. J.)

FIG. 594
A typical English double-decked bus which is growing in popularity in this country.

in England, where hundreds have to be used in the city because no trolley lines or elevated lines are permitted. These buses are growing in popularity all over the country, as in clear weather the upper deck affords a fine view of the city and streets it traverses and gives the passengers a bracing open-air ride. Fig. 594 shows a typical bus of this type.

The jitney bus is a development of recent months rather than of years and from a vehicle, automobile or body standpoint has no set canons. A jitney bus can be and is any vehicle that will run on wheels and carry four, five or more people. The touring car used for this purpose cannot compete with a speed wagon chassis with special passenger body on it because the latter can carry so many more people each trip.

FIG. 595
Typical bus construction for short trips. (Vehisote panels used).

Buses used in this business are made as simple and cheap as possible because competition is so keen that expensive equipment cannot be afforded. A typical bus construction suitable for jitney service, bus service, or short country trips is shown in Fig. 595. The buses shown in Figs. 596 and 597 are special ones developed for hotel and community service. In Fig. 596 the bus is equipped with luxurious upholstery and has striking lines, while the bus shown in Fig. 597 is well laid out for service and provides plenty of accommodation, fine large windows and protection against any kind of weather.

An American army truck body is shown in Fig. 598, This body is very substantially constructed and is really of the platform type, because the sides are detachable from the base or platform. The entire body is made so that it can be knocked down, and packed into a compact unit. The sides are high and give good carrying capacity; for weather protection hoop bows are used with a cloth covering, giving somewhat of an effect of a prairie schooner when the covering is on the top. For army or any field work a truck body of this design is the most practicable of any that have been developed.

FIG. 596
Hotel bus on striking lines and luxuriously equipped.
(Collings Carriage Company, Camden, N. J.)

Fig. 597

A 3½-ton Rowe operating of regular schedule between several New Jersey towns. The pneumatic tires make it as easy riding as a Pullman. The type is well adapted for community service.

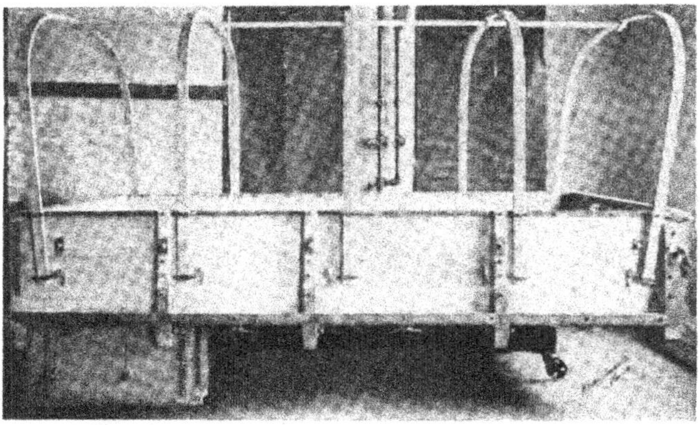

FIG. 598
An American army truck body, "knock-down" construction.

FIG. 599
"Everyway" body, adapted to carrying live stock.

FIG. 600
"Everyway" body, adapted for carrying grain or other loose material.

FIG. 601
"Everyway" body with sides let down at an angle, for carrying sacks and bales.

FIG. 602
"Everyway" body; sides at right angle to body, for carrying boxes and crates.

FIG. 603
"Everyway" body, adapted for carrying loose or baled hay.

FIG. 604
Contractors' dump body with mechanical hoist using engine's power.

FIG. 605
Examples of all steel heavy dumping bodies with hoisting equipment.

There are big possibilities ahead for the motor truck designer and builder to develop equipment for the farms. There has been some movement in this direction but there are vast opportunities in this field. One large company has developed a farm wagon which has a wide utility. Most of the equipment developed for farm service and transportation has been along the lines of special bodies on standard trucks. One of the most adaptable forms of bodies developed is shown in Fig. 599. The manufacturer calls this the "Everyway" body. The

Fig. 606
A unique feature of this White 3-ton truck of the Burton-Dixie Corporation, Brooklyn, N. Y., manufacturers of fabricated cotton batts, is the sleeping chamber built over the driver's compartment. On long trips the sleeping chamber has been a saver of both time and money. It enables two drivers, by alternately driving and resting to keep the truck traveling all the time.

body shown in Fig. 599 is adapted for carrying stock. The stakes are 60 inches high and have a rope threaded through loops in their ends to prevent the cattle from attempting to climb out over the sides. In Fig. 600, the sides of the body are filled up with slats so they make a high grain-tight body that can be used for handling grain or any other loose material.

In Fig. 601 is shown the same combination as Fig. 600, only the side brackets are adjusted to let the sides down so that they make an angle of 30 degrees, making the body adaptable for handling sacks, bales of hay, etc. Grain can also be hauled to the extent of the side height. When the side brackets are dropped

Fig. 607
A good example of the modern type of furniture moving body.

FIG. 608
United States army ambulance on a light truck chassis.

FIG. 609
United States army ambulance body, showing interior arranged for sitting and for prone patients. The upper picture illustrates the ambulance in actual service.

down as shown in Fig. 602 and the sides set at a right angle to the body they adapt it for carrying passengers or for hauling crated and boxed fruit and vegetables. The posts and back rail can be removed as shown in Fig. 603; then the body can be used for carrying loose hay. The Comstock gate makes it suitable for heavy, loose grain. The "Everyway" body is very cleverly worked out and should appeal to the farmer who is favorable to combination equipment. It also furnishes food for thought to the body designer.

The special body equipments for commercial and general utility, like contracting and city use, convey a big field. A few bodies are shown in Figs. 604 to 609. These give the readers some idea of the possibilities in truck body design.

FIG. 610
Truck trailer of the platform type with side stakes.

In the heavy truck field a dumping body is perhaps the most used and typical examples of these bodies as are shown in Figs. 604 and 605. All steel construction prevails on this class of body. A novel body is shown in Fig. 606. This, beside being very large, has a sleeping compartment on top of the driver's cab. Moving wagons are large closed bodies. A good example of a furniture moving body is shown in Fig. 607. Truck chassis have been used extensively for ambulances and typical examples of this type of wagon are shown in Figs. 608 and 609.

The truck trailer is used a good deal and deserves some attention because of the possibilities there are of its extensive use. The body equipment is generally very simple and is of the platform type with side stakes, similar to that shown in Fig. 610, but there is no limit to the construction used. There is a company in Detroit which has a large trailer with canvas covered body that is used to haul closed bodies to nearby cities. The trailer is especially adaptable for hauling bulky material of light weight as the amount carried can be doubled for the same running expenses. Time is saved and also one driver can control the truck and trailer, whereas, if two trucks were used two drivers would be required.

TRUCK CABS

Motor truck development has been so rapid that many things have been in a measure overlooked. The touring car took a long time before it arrived at its present state and similarly the truck has to be developed and perfected step by step. One of the things that truck development is now demanding is the better protection and comfort of the driver.

The driver's seat and room have been cut down to the minimum in order to allow for a maximum loading space. When the seat and leg room are crowded up and improperly shaped seats and cushions are used, drivers cannot be expected to operate their trucks as efficiently as if they would if they had ample accommo-

FIG. 611
The latest trucks have full spring cushioned seats and backs. This is a good selling feature.

dations. The latest trucks developed have full spring cushioned seats and backs like that shown in Fig. 611. This is a good selling feature because the driver, if he were consulted, would quickly choose the truck which afforded him the most comfort. The old standard shape for seat back consists of a board with a hair or cotton pad covered with leather.

If the truck is to be operated in all weathers some means must be provided to protect the driver. Collapsible fabric tops and fabric side curtains are but makeshifts, and in constant service soon become torn. A truck for all around service should have a closed compartment for the driver. This compartment can be made in several ways and has to be arranged to match up the body that is to be used. When the body has a canopy top the compartment has to be built into it. If a panel body is used it is only necessary to fit full doors at the front to enclose the cab.

FIG. 612
Same seat as in Fig. 611, but shown in two parts, the lower section making a good tool box.

The seat construction shown in Fig. 611 is designed for quantity production and also to permit use of the truck with or without a cab. In fact it can be used for canopy top bodies, open express bodies without a top or with a fabric passenger car type top, or a top like that shown in Fig. 613. In Fig. 612 is

FIG. 613
Driver's cab, with the seat shown above in position.

shown the seat in two parts, the lower section, which is the support, makes a good tool compartment.

A full back is used, with springs, and a high seat cushion made in two sections. This makes a very comfortable seat arrangement. In Fig. 613 is shown how the seat is adapted by bolting on sides and a top making a very neat open type cab. The substantial windshield construction should be noted. The shield sways both inward and outward. For bad weather side curtains have to be used. The commercial possibilities in a design of this kind are shown in Fig. 614, where all main parts are shown dismantled ready for crating in the box, the parts of which are laid around it. In designing body equipment the knock-down feature is a very important one, as it makes the freight rates lower and is more convenient all around.

The ideal truck condition provides a cab or compartment for the driver so he is protected under all conditions. The wide variety of trucks used calls for a more effective protection for the driver than fabric curtains can give. Even for short hauls the cab with properly designed doors will give better efficiency, as

FIG. 614
Cab with all the main parts dismantled ready for crating. The knock-down feature is a very important one, as it reduces freight rates and is more convenient all around.

fabric curtains with celluloid soon get torn with the rough usage the average truck driver will give them. The cab shown in Fig. 615 is the next step in the combination of seat shown in Fig. 611. One side of this cab is omitted to show the interior. The completed cab is shown in Fig. 616. It will be noted that a large windshield and liberal sized lights are provided in the doors and sides next to the driver's seat.

There is a big market for truck cabs which are economically but strongly made so that they will give the maximum service for the least cost. The cab shown in Fig. 616 is also shown in Fig. 617, but mounted on a chassis. A truck chassis like that shown in Fig. 617 has a wide market because any type of body can be fitted behind the cab. A larger cab is shown in Fig. 618, mounted on a chassis with a platform and stake body behind it. The combination will be seen to work out very well. A well-developed all-steel cab is shown in Fig. 619. A large adjustable windshield and liberal sized quarter lights are used, with deep cushions and spring back heavily padded. The cab shows that considerable attention has been paid to its details, as it shows a very clean cut finish.

Fig. 615

Fig. 616

Fig. 617

Figs. 615, 616 and 617 show further steps in the development of the seat in Fig. 611; in Fig. 617 is shown the completed vehicle.

The possibilities of interchangeable manufacturing of truck cab bodies and parts is exemplified in Figs. 620, 621 and 622. In Fig. 620 is shown a small panel body on a Ford. The front end or driver's compartment is made from the same parts as shown in the cab in Fig. 619, the rear end of the body being the

FIG. 618

FIG. 619

FIG. 618
Large cab with platform and stake body behind it.

FIG. 619
A well developed all-steel cab in which much attention has been paid to details resulting in a clean cut job.

FIG. 620
Panel body on Ford chassis with cab made from same parts as shown in Fig. 619.

FIG. 620

only part needed. In Fig. 621 a canopy top body is shown. This body used the same parts of the cab as Fig. 620, only the roof is replaced by the canopy top extension. This makes a very practical combination. In Fig. 622 is shown an

FIG. 621
Canopy top body using same parts as in Fig. 620, only the roof is replaced by the extension of the canopy top.

FIG. 622
Open express body with cab attached. The outer side extends beyond the body to make a running board and fender. The brace system insures very strong side panels, and the body width is the same as the cab.

open express body with the cab attached. The body construction is novel because the outer sill extends out from the body to make a running board and fender. The brace system insures very strong side panels. The body width is the same as the cab and the whole makes a very neat design.

In hot weather the side doors can be removed or fastened back against the sides of the body. The windows in the rear sides, by the seat, can be either removed or stored in racks in the cab, or dropped down, depending on the construction and design of the cab. When the design arranges for these conditions the cab is very comfortable in the summer. The cab shown in Fig. 616 has a small hand hole for signalling. This is covered by a slide.

FIG. 623

Sliding door cab showing the cab wide open on the left and closed on the right. (Highland Body Mfg. Co., Cincinnati, Ohio.)

An improved cab, of wood and steel construction, is shown in Fig. 623. This cab has sliding doors, which slide into pockets and automatically lock in either open or closed position, while the windows slide and fold into the back of the cab, and are locked in all positions. The rear window is made to drop. These cabs are strong, yet light in weight.

APPENDIX

Comprising some useful data for Automobile Body Designers and Builders.

A PERFECTED "CONTRACT FORM" FOR THE USE OF AUTOMOBILE BODY BUILDERS

The following tabulation is published because it is the most complete listing of the parts, units, materials, etc., entering into the body construction that has yet come to our attention. As used by its originators, the list is printed on a series of ruled sheets enclosed in an appropriate folder. Preceding this list there are blank spaces for name of customer, dates, chassis data, etc. On the folder terms and conditions are printed. Spread out, with ample space for writing in the specifications, the list printed below occupies nine sheets, 8½ by 11 inches. This contract form was originated and is used by Smith-Springfield Body Corporation, Springfield, Mass., and New York City.

BODY CONSTRUCTION

Frame
Frame Ironing
Panels
Mouldings
Seats
  Front (Height
    (Width
    (Depth
  Rear (Height
    (Width
    (Depth
  Aux.
Doors
Roof
  Material
  Exposed or Concealed Bows
Glass
  Doors (
    (
  Rear Quarter (
    (
  Back (
    (
  Division (
    (
  Windshield
  Visor

Body Hardware Continued—

Heater
Roberail
Door Strikers
Anti-rattlers
Door Bumpers
Instrument Board
Ventilators
  (Top Cowl
  (Side Cowl
  (Roof
Telephone
  Telephone Face Plate
Running Board Mouldings
Floor Board Mouldings (Front
  (Rear
Scuff Plates
Floor Board Lift Plates
Door Fixtures for Folding Seats

EQUIPMENT

Radiator
Hood
Front Fenders

PAINT

Color (Below Belt
Belt Panel
Upper Panel
Top of Cowl and Hood
Fenders
Radiator
Roof
Chassis
Splashers
Wheels
  Mouldings
  Hollow Ledges
  Sash Mouldings
Stripe
Front Seat Interior
W. W. Above Belt
Monogram

TRIMMING MATERIAL
(REAR COMPARTMENT)

Cloth (Body
  (Headlining
Lace (Pasting
  (Seaming
  (Broad

Style of Trim, Rear, (Continued)—

Side Quarters and Division
  Plain
  Paneled
Headlining
  Roof
  Sides
Pockets (Misc.)
Door Stops

TRIMMING MATERIAL
(FRONT COMPARTMENT)

Cloth (Body
  (Headlining
Leather (Color
  (Grain
Lace (Pasting
  (Seaming
  (Broad
Curtain Cord
  Sash Silk
Curtain Rollers
Curtain Tassels
Nails (Finishing
Carpet

BODY HARDWARE

- [illegible]
  - Door
  - Pillar
  - Hood
  - Tool Box
  - Battery Box
  - Tire
  - Cabinet
  - Top
  - Trunk
  - Floor Board
- Hinges
  - Door (Exposed (Concealed
  - Butts
  - Hood
  - Cabinet
  - Top
- Regulators
- Window [illegible]
- Handles
  - [illegible] [illegible] ((
  - Inside [illegible]
  - Pull to
- Grab
- Pillar
- Regulator
- Hood
- Seat
- Cabinet
- Lights
  - Dome
  - Corner
  - Bow
  - Step
  - Dash
  - Head
  - Tail
    - [illegible]
  - Pillar
    - [illegible]
- Rear Fenders
- Splashers
  - (Front
  - (Rear
  - (Side
- Running Boards
- Steps
- Brackets
  - ([illegible]
  - (Step
  - (R-Board
  - (Fender
- Lights
  - Head
  - Cowl
  - Pillar
  - Tail
- Tire Carrier
- Battery Box
- Trunk
- Trunk Rack
- Luggage Rack
- Gas Tank
- Bumpers (Chassis)
- Rear [illegible] Mirrors
- Switches
- Switch Plates
- Cigar Lighters
- Footman Loops
  - [illegible] [illegible]
- Watch

TOP

- Bows
- Irons
- Rests
- Material
- Curtains
- Curtain Lights (Glass (Celluloid
- Headlining
- Envelope
  - [illegible] Irons
- Curtain Cord
- Curtain Silk
- Curtain Rollers
- Curtain Tassels
- Nails [illegible]
- Carpet
- Carpet Binding
- Rubber Mat
- Thread
- Arm Rests
- Arm Straps
- Foot Hassocks
- Foot Rest

STYLE OF TRIM
(REAR COMPARTMENT)

- Body
- Cushion
- Divided
- Pleated
- Tufted (1 Row Buttons (2 [illegible] Buttons
- Diamond
  - (1 [illegible] [illegible]
  - (2 [illegible] [illegible]
  - Plain
- Back
  - Divided
  - Pleated
  - Tufted
    - (1 Row Buttons
    - (2 Row Buttons
  - Diamond
    - (1 Row Bu[illegible]
    - (2 Row Bu[illegible]
- Back
  - Plain
- Doors
  - Plain
  - Paneled
  - Pockets
- Carpet Binding
- Rubber Mat
- Thread
- Arm Rests
- Arm Straps
- Foot Rest

STYLE OF TRIM
(FRONT COMPARTMENT)

- Cushion
  - Divided
  - Pleated
  - Tufted
    - (1 Row Buttons
    - (2 Row Buttons
  - Diamond
    - (1 Row Buttons
    - (2 Row Buttons
  - Plain
- Back
  - [illegible]
  - Pleated
  - Tufted
    - (1 Row Buttons
    - (2 [illegible] Buttons
  - Diamond
    - (1 [illegible] Buttons
    - (2 Row Buttons
  - Plain
- Doors
  - Plain
  - Paneled
  - Pockets
- Side Quarters
  - Plain
  - Paneled
- Headlining
  - Roof
  - Sides
- Pockets (Misc.)
- Door Stops

COMPARISON OF DIFFERENT TYPES OF GLUE

Table prepared by U. S. Forest Products Laboratory

Particular compared	Animal glue	Casein glue	Vegetable glue	Blood glue	Liquid glue
Source	Hides,bones, horns, etc.	Casein from milk	Cassava starch	Dried Blood	Animal glue or fish parts
Cost per lb. 1920	25-42 cents	16-20 cents	10-12 cents	20 cents	$1-$5 per gal.
Spread in sq.ft.per lb.	25-35	35-55	35-50	30-100	No data
How mixed	Soaked in water and melted	Mixed cold with rapid stirring	Mixed with alkali and cold or hot water	Mixed cold	No preparation
How Applied	Warm with brush or mechanical spreader	Cold with brush or mechanical spreader	Cold with mechanical spreader not by hand	Cold with brush or mechanical spreader	Cold or warm usually applied by hand
Temperature of press	Cold, or with hot cauls	Cold	Cold	Hot	Cold
Strength (in shear test)	High grades stronger than strongest woods	Equal to medium grade animal glue	Equal to medium grade animal glue	High strength in plywood Not used for joint work	Best grades equal to medium grade animal glue
Water resistance	Low	High	Low	High	Low
Chief uses in woodworking	For strong joint work	For water resistant plywood or joint work	For veneer work because of cheapness	For water resistant veneers	For repair work and small articles

A useful table showing the Properties of Glues. Prepared by Forest Products Laboratory.

WHEN SHOULD LUMBER BE TAKEN FROM THE KILN?

The two things which a dry kiln operator must be able to prove at the end of a kiln run are that his lumber is as dry as is required and that it is free from invisible seasoning stresses which would cause warping when the wood is resawed and shaped into body parts or other products.

Neither of these points can be proved by examination of the outside of the lumber. They are easily determined, however, by the following tests which have been devised by the U. S. Forest Products Laboratory, Madison, Wisconsin.

Before the lumber is removed from the kiln, choose a fairly representative board from each truck load of stock. Cut four 1-inch cross-sections (A, B, C, D) from the board at least 2 feet from the end.

Use section A to find the average moisture content of the dried stock. To do this, weigh the section immediately after cutting, on a balance accurate to one-tenth of one per cent., and then dry it on a steam pipe or in an oven 212 degrees F. until it reaches constant weight. The weight lost during this drying is the weight of the moisture which was in the section. Divide the weight of the moisture by the weigh of the oven-dry section and multiply by 100. This will give the percentage of moisture in the section and also in the stock in that part of the kiln from which the sample was taken.

For high grade uses, the moisture content of any board in the kiln should not vary by more than 2 or 3 per cent. from the final moisture content specified.

Use section B to find out whether the stock is uniformly dry from center to outside. In order to do this the section must be cut apart and the moisture content of the inside and outside found separately. If the stock is 1½ inches or more in thickness, cut the section parallel to its edges to get an outer shell of material ¼ inch thick.

Trim the remaining block equally on all four sides to leave a core one-half inch thick. If the stock is less than 1½ inches thick, cut section B so as to get an outer shell and inner core each one-fifth the total thickness of the section. Find the moisture content of each piece by the method used for finding the moisture content of Section A.

The moisture content of the inside and outside of the stock should be equalized by steaming, if necessary, to within two per cent. before the lumber leaves the kiln.

The third and fourth sections, C and D, are for casehardening and moisture distribution tests. Saw section C parallel to the wide faces of the original board to form tongues or prongs, leaving about one-half inch of solid wood at one end of the section. If the stock is less than two inches thick, make two saw cuts; if it is more than two inches thick, make five saw cuts.

From sections having six prongs break out the second prong from each side, leaving two outer and two central prongs. From sections having only three

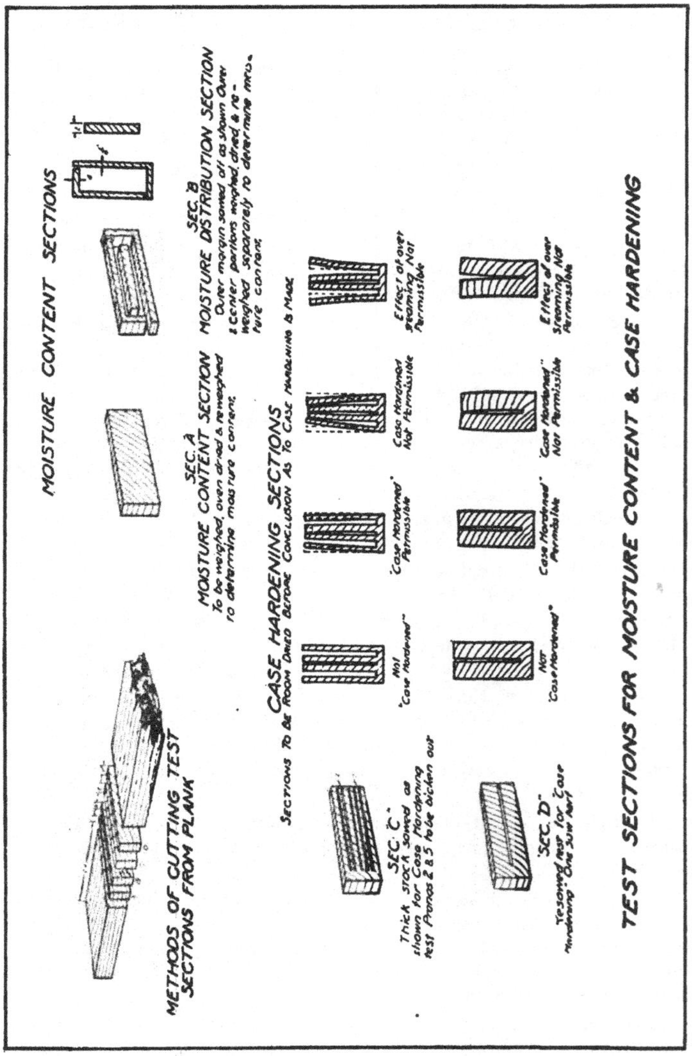

TEST SECTIONS FOR MOISTURE CONTENT & CASE HARDENING

prongs remove the center prong. In section D saw one central saw kerf to form two prongs. Stand the sections on end in some convenient place in the shop to dry.

Observe carefully the action of the prongs from the moment of sawing. Do they bow in or out or remain straight on the saw? Do they change shape after room drying?

If the prongs remain straight both on the saw and after room drying, the lumber is perfectly seasoned, being free from stresses and uniformly dry throughout.

If the prongs remain straight on the saw but turn in after room drying, the moisture distribution is uneven, the surface being drier than the inside. A short steaming treatment to balance the moisture content should relieve all stresses.

If the prongs turn in on the saw and do not turn out after room drying, the lumber is "casehardened" and is drier outside than inside. Use a steaming or high humidity treatment to moisten and soften the surface. The shrinkage of the outside in redrying should relieve the interior tension and cause the stresses to disappear.

LUMBER USED IN THE MOTOR VEHICLE INDUSTRY

The rapid increase in the proportion of closed cars manufactured is an outstanding feature of the automobile industry. An official of a large company recently expressed his belief that in five years one-half of their output would be closed cars. Already one out of every eight pleasure cars of a well-known make is a sedan or coupé. This means a large consumption of lumber, as the closed car takes from two to three times as much lumber as an open car, and a better grade of lumber is required to insure rigidity and freedom from warping in the closed body. An engineer of the U. S. Forest Products Laboratory, Madison, Wisconsin, recently visited a number of manufacturing plants to determine what woods are now being used in the automobile industry, to what extent substitution of one species for another is taking place, and what troubles manufacturers are having

Table I.—Kinds of Wood Used in Open Cars.

Sills, longitudinal and cross	Ash, hard maple and occasionally elm, red gum, magnolia and soft maple
Floor boards	Sound and wormy oak, hard and soft maple, red gum, beech, wormy chestnut, elm
Seat risers, or "heel boards"	Hard and soft maple, red gum, yellow pine
Seat boards, or seat frame	Hard and soft maple, red gum and numerous other species
Seat lids	Maple, gum, elm and numerous other species and plywood
Pillars and posts	Hard and soft maple, ash, elm, sycamore and red gum
Seat rails (arm and back)	Ash, elm and maple
Strainer slats, or "spring slats"	Maple, ash and gum
Doors	Hard and soft maple, ash and elm
Trim rails	Rock elm
Running boards	Wormy oak, yellow pine, maple, Douglas fir
Steering wheels	Walnut, maple, red gum
Spokes	Hickory
Rims ("felloes")	Hickory
Top bows	Oak
Dash	Cottonwood and maple

with wood. He found that maple leads for use in the construction of bodies, elm is next, and ash is third. More 2 and 3-inch stock is used than thinner materials. The following comments are to be noted in connection with the use of various woods.

Maple—At most of the plants visited by the representative of the Forest Products Laboratory, maple was used for body sills, (in one plant practically the entire framework and even the floor boards of the car were made of maple), although ash is used for sills at some of the plants. Maple is cheaper and is generally more uniform in quality than ash, and warps less than elm. In one plant birch was suggested as a substitute for maple, but it is more expensive. Maple is said to hold screws less rigidly than elm because it is less fibrous and after use the screw hole becomes enlarged and smooth, permitting the screw to come out easily. One company preferred birch to maple in sedans because they claimed it would take and hold varnish better, especially on rounded corners. Maple was also said to split more easily than elm, making it difficult to saw curves with economy.

Elm—seems to be the principal wood used for the framework of open bodies. Soft elm is used except for the trim rails, which are of rock elm or a good grade second-growth soft elm bent to proper shape. Soft elm works easily, holds screws well, and does not split easily. Stock up to 4 inches in thickness is used.

Table II.—Amount of Lumber Used Annually in the Manufacture and Shipment of Passenger Cars and Motor Trucks. Based on 1919 Production.

Total output of passenger cars	1,660,000	
Average number of board feet of lumber used per car	160	
Total lumber used in passenger cars		265,600,000 bd. ft.
Total output of motor trucks	316,500	
Average number of board feet of lumber used per truck, including body	200	
Total lumber used in motor trucks		63,300,000 bd. ft.
Total number of passenger cars exported	66,400	
Average number of board feet of lumber used in export crating of a passenger car	660	
Total lumber used in export crating of passenger cars		43,824,000 bd. ft.
Total number of trucks exported	15,825	
Average number of board feet of lumber used in export crating of trucks	760	
Total lumber used in export crating of motor trucks		12,027,000 bd. ft.
Grand total		384,751,000 bd. ft.

Ash—Ash was formerly used almost exclusively in automobile work, but due to its greatly increased cost it is now used only on high-priced cars, and on cars with closed bodies which command a relatively high price. Ash is preferable for use in framework of closed cars because it holds its shape well. At one plant trouble was experienced with maple as compared with ash, and this made the cost of maple equivalent to that of ash. A tough sill is required to reinforce the steel frame. That the wood actually reinforces the steel frame is shown by the fact that breaks in the frame usually occur at the front end of the wood sills, that is, near the dash.

Birch—The use of birch is probably increasing in automobile manufacture. Some manufacturers report its use in sills and frames. It is preferred to maple on exposed painted parts because it is said to hold the paint better. It is said to be as good as maple as far as mechanical properties are concerned and better in seasoning properties, but usually it is slightly more expensive.

Hickory—This species is used for spokes and rims only.

Gum—An appreciable quantity of red gum is used for various parts, such as foot risers, foot boards, strainer slats and floor boards. It is also being used for frames and steering wheels.

Other Species—Among the other species used for minor parts of automobiles are the following: Wormy oak is used for running boards, floor boards and foot risers. For top bows second-growth, at least so-called second-growth oak, is used principally, although some elm is now used. Sycamore is used to some extent for posts and pillars. Yellow pine is used for floor boards and running boards. Douglas fir has been used for the same purpose as yellow pine. Basswood, cottonwood and yellow poplar are used for minor parts.

Table 1 shows the kinds of wood used in the different parts of an automobile, based on a study of these companies which made bodies for a number of automobile manufacturers.

It is estimated by the Forest Products Laboratory that the total amount of wood used in the construction of automobiles and motor trucks in the United States amounted to 384,751,000 feet B. M. in the year 1919. The total consumption of wood used in the industry is roughly estimated in the accompanying Table 2. (See Table 2.)

The amount of lumber used in each car varies from 75 feet B. M. for a small open car to 200 feet B. M. for a medium-priced touring car. An average given by a large body manufacturing corporation is 140 to 150 feet B. M. for open cars for each body. A small sedan requires 225 feet B. M. and a large sedan, not including running boards. uses about 310 feet B. M. One company stated that the average waste was about 30 per cent., including drying losses, cutting and minimum jointer waste, although others place the waste as high as 40 per cent.

In automobile work first and seconds are used nearly exclusively. One company used 75 per cent. first and seconds and 25 per cent. No. 1 common. A large body company used 40 to 50 per cent. first and seconds, and the rest No. 1 common of maple, elm and oak. Another company making high-priced cars will take only 20 per cent. of No. 1 common.

SUBSTITUTES FOR ASH IN AUTOMOBILE BODIES

Ash has always been considered the most desirable wood for use in automobile bodies. It combines the properties of moderate weight, easy workability, high degree of toughness and comparative freedom from warping. On account of the high price of ash, however, other woods are gradually replacing it in many of the less expensive cars.

The following description prepared by the Forest Products Laboratory gives some of the advantages and disadvantages of the substitute woods as compared with forest grown ash for automobile construction:

MAPLE. Hard maple is used for sills in many cars, and in some for the framework of the body and even the floor and running boards. Maple is fully as strong and stiff as a beam or post as white ash, but is not as shock-resistant. It is usually cheaper than ash and runs more uniform in strength. Maple warps very little, in this respect being superior to elm. On the other hand, maple is more difficult to season without checking than ash or elm, and it is said not to hold screws so well in motor car bodies. On account of the smooth, fine texture of maple, paint and enamel rub off it more easily, especially on carved surfaces which receive considerable wear, than off birch, which is slightly more porous. Because of its smooth-wearing qualities and comparative freedom from slivers, maple is preferred to all other woods for the floors of delivery trucks.

ELM. The principal use of elm is for frames, seat backs and doors; very little, if any, is used for sills. White is preferred to rock elm, except for some of the bent parts, because it is more easily worked and is less subject to warping. For the same reasons lumber from old white elm trees, usually called "gray elm," is preferred to that from younger or vigorously growing trees. Old white elm is not so strong or tough as ash, on the average, but it varies less in strength than ash, especially that which comes from the Southern swamps.

BIRCH. Yellow birch is a close rival of maple. It is used for sills, framework and many minor parts. It is said to hold the paint better than maple on exposed parts.

HICKORY. The true hickories are used almost exclusively for spokes and felloes. Te pecan hickories, which are somewhat inferior as a class to the true hickories, might be used in body construction, although their hardness and tendency to twist would perhaps prove a serious drawback.

RED GUM. Red gum is too weak and soft for the sills and other major parts of the frame, but is used for floor boards, seat risers and other minor parts. One of the principal drawbacks to the use of gum is its tendency to warp with changes in moisture content. Quarter-sawed gum gives less trouble in warping than plain-sawed gum.

OAK. In automobile construction no distinction is made, as a rule, between the different species of oak or even between the red oak and white oak groups. In truck bodies, oak is one of the leading woods, being used for sills, cross sills, frames, floors and stakes. In pleasure cars oak is rarely used for the frame or sills.

Wormy oak is used for running boards, floor boards and seats and some sound oak for instrument boards and battery boxes. Top bows are made almost exclusively of oak, second growth being preferred.

SOUTHERN YELLOW PINE. Under this heading are included longleaf, loblolly, shortleaf and some of the minor Southern pines. These have been found adaptable for running boards, floor boards, seat boards and a number of small parts in the seats and frames.

OTHER SPECIES. Cottonwood is used for dash boards of pleasure cars and the boxes or bodies of trucks. Sycamore, beech, basswood, yellow poplar, cucumber, tupelo, gum, chestnut, Douglas fir and western yellow pine have also entered into car body construction to a small extent.

The comparative merits of the different species in the four properties most important in automobile construction are given in the following table, the strength of forest-grown white ash being taken as 100. Actual strength values of these species are to be found in Department of Agriculture Bulletin 556, "Mechanical Properties of Woods Grown in the United States."

Strength of Woods Used in Automobile Construction in Per Cent. of the Strength of Forest-Grown White Ash.

HARDWOODS Species	Strength as a beam or post	Stiffness	Shock resisting ability	Hardness
Ash, white, forest grown	100.0	100.0	100.0	100.0
Ash, black	71.3	79.3	90.1	62.3
Ash, white, second growth	122.5	117.6	119.6	118.9
Basswood	59.1	80.6	40.5	29.6
Beech	93.5	96.9	96.0	90.0
Birch, yellow	104.8	116.8	120.6	80.9
Chestnut	66.0	71.9	53.4	49.2
Cottonwood	60.6	79.0	54.3	35.3
Cucumber	85.4	112.4	76.7	54.9
Elm, rock or cork	98.8	92.9	140.5	101.6
Elm, white	79.2	79.5	89.5	57.1
Gum, red	80.7	91.5	75.5	59.0
Gum, tupelo or cotton	81.4	82.5	63.5	77.3
Hickories, pecan	103.5	103.8	119.7	139.6
Hickories, true	126.6	120.2	173.9	150.4
Maple, red	90.0	101.2	78.7	75.4
Maple, silver	66.9	68.5	71.7	64.3
Maple, sugar	104.7	105.9	90.5	103.0
Oaks, all kinds	92.6	101.3	94.9	104.5
Poplar, yellow	67.3	93.8	41.5	37.9
CONIFERS				
Fir, Douglas, Pacific Coast	95.7	122.1	59.9	58.3
Pine, loblolly	93.7	105.6	71.0	60.0
Pine, longleaf	112.2	122.1	77.7	74.8
Pine, shortleaf	94.1	100.6	69.7	64.0
Pine, western white	75.5	99.7	53.8	37.0
Pine, western yellow	67.0	75.6	42.9	41.0
Spruce, Sitka	69.5	94.1	63.3	44.9

STRENGTH OF SCREW FASTENINGS IN PLYWOOD

If the screw fastenings in plywood construction are to be as strong as the plywood itself, it is important to adapt the size of screw, spacing and margin to the particular species and thickness of plywood used. Tests made at the Forest Products Laboratory have shown that the commonly-used plywood species may be divided into the following groups, all woods in any one group requiring the same screw fastening to develop maximum strength.

GROUP I *Low Density*		GROUP II *Medium Density*		GROUP III *High Density*
Basswood	Hemlock	Ash, black	Hackberry	Ash, white
Cedar, Spanish	Pine, sugar	Ash, pumpkin	Magnolia	Beech
Cottonwood	Pine, white	Elm, white	Mahogany	Birch
Cypress, bald	Poplar, yellow	Gum, black	Maple, soft	Cherry, black
Douglas fir	Redwood	Gum, cotton	Sycamore	Elm, cork
Fir, true	Spruce, Sitka	Gum, red	Walnut, black	Maple, hard

SIZE AND SPACING OF SCREWS FOR MAXIMUM STRENGTH IN PLYWOOD

Species of plywood	Thickness of plywood in inches	Gauge (number) of screw	Screw length in inches Species receiving point: *White ash*	*Spruce*	Margin in inches	Spacing in inches
	3/30	4	1/2	5/8	1/2	3/8
	3/24	5	1/2	5/8	5/8	1/2
	3/20	6	5/8	3/4	5/8	1/2
GROUP I	3/16	7	5/8	3/4	5/8	5/8
	3/10	9	3/4	1	3/4	3/4
	3/8	11	1	1 1/4	3/4	3/4
	3/30	5	1/2	5/8	1/2	1/2
	3/24	6	5/8	3/4	5/8	1/2
	3/20	7	3/4	7/8	5/8	5/8
GROUP II	3/16	8	7/8	1	5/8	3/4
	3/10	10	1	1 1/4	3/4	3/4
	3/8	12	1 1/4	1 1/2	3/4	7/8
	3/30	6	5/8	3/4	1/2	1/2
	3/24	7	3/4	1	5/8	5/8
	3/20	8	1	1 1/4	5/8	3/4
GROUP III	3/16	9	1 1/4	1 1/2	5/8	3/4
	3/10	11	1 1/2	1 3/4	3/4	7/8
	3/8	13	1 3/4	2	3/4	1

The screw sizes, margin and spacing for use with each species and plywood thickness will be found in the accompanying table. The gauge is the smallest that can be used with the thickness specified and not cause failure through breaking of the screw when the full strength of the plywood is developed. The length of screw is the shortest which will prevent the screw from pulling out before the full plywood strength is reached. The margin is the smallest distance from edge of hole to edge of plywood which will insure against failure by shear. The spacing is the distance from center to center of screw holes which gives maximum strength per linear inch.

About equally good results were obtained with flat-headed screws without washers, and round-headed screws with washers. Round-headed screws without washers proved an inferior means of fastening. The spacing given in the table is for screws in a single row, but staggering is recommended wherever possible.

In the tests the size of frame members to which the plywood might be attached was necessarily a secondary consideration, and the block of wood in which the screw points were held was simply made large enough to prevent failure occurring in it. Until further information is obtained, designers must take particular care that the frame is not split or weakened through the use of the size of screw and the spacing necessary to make the fastening as strong as the plywood.

USEFUL TABLES OF WEIGHTS FOR TRUCK DESIGNERS

BULK FREIGHT

	Wt. Lbs.
Asbestos, per cu. ft.	192
Asphaltum, per cu. ft.	87
Ashes, of soft coal (solid), per cu. ft.	43
Bricks, per 1,000	4,500
Bricks, soft, per cu. ft.	100
Brick, common, per cu. ft.	112
Brick, hard, per cu. ft.	125
Brick, pressed, per cu. ft.	135
Brick, fire, per cu. ft.	140-50
Brick, sand lime, per cu. ft.	136
26 ordinary bricks equal 1 cubic foot	118
Clay, per cu. ft.	120-150
Clay, per cu. yd.	3,240-4,050
Cement (Portland), per barrel	380
Cement (Portland), per bag	94
Cement (Natural), per barrel	285
Cement (Natural), per bag	94
Concrete, per cu. ft.	120-155
Earth (rammed), per cu. ft., equals	110
Earth (rammed), per cu. yd., equals	2,970
Emery, per cu. ft.	250
Fuel Oil, per cu. ft.	54
Gravel, per cu. ft., equals	80-110
Gravel, per cu. yd., equals	2,160-2,970
Gold or Silver ore, per cu. ft.	154
Gold or silver ore, per cu. yd.	4,158
Granite, per cu. ft.	160-170
Grain, wheat, per bushel	60
Grain, oats, per bushel	30
Grain, corn, per bushel	56
Grain, rye, per bushel	60
Garbage, per cu. ft.	50
Hay, baled, per cu. ft.	95
Hay, pressed, per cu. ft.	25
Iron, cast, per cu. ft.	446
Iron, grey foundry, per cu. ft.	450
Lime, per struck bushel	75
Lime, per cu. ft.	55
Mortar, per cu. ft.	110
Mud, per cu. ft.	102
Marble, Italian, per cu. ft.	169
Marble, Vermont, per cu. ft.	165
Oil, petroleum, per cu. ft.	55
Plaster of Paris, per cu. ft.	93-113

BULK FREIGHT (Continued)

	Wt. Lbs.
Sand, per cu. ft.	110
Sand, per cu. yd.	2,970
Salt (Syracuse), per struck bushel	56
Salt (Turks Island) per struck bushel	80
Stone (various), per cu. ft.	175
Stone (various), per cu. yd.	4,725
Sand stone, per cu. ft.	155
Steel, per cu. ft.	490
Slate, per cu. ft.	175
Tar, per cu. ft.	62
Water, salt, per cu. ft.	64
Water, rain, per cu. ft.	62

ARTIFICIAL ICE

	Wt. Lbs.
4 x 10 x 24	25
6 x 12 x 26	50
8 x 15 x 32	100
11 x 11 x 32	100
8 x 15 x 44	150
10 x 15 x 36	150
10 x 20 x 36	200
11 x 22 x 32	200
14 x 14 x 40	200
11 x 22 x 44	300
11 x 22 x 56	400

COAL

	Wt. Lbs.
1 bushel (Bitum. coal), 1.55 cu. ft. equals	75
Anthracite coal, 1 cu. ft. equals	66
Bituminous coal, 1 cu. ft. equals	55
Cannel coal, 1 cu. ft. equals	50
Charcoal, 1 cu. ft. equals	20
Coke, 1 cu. ft. equals	32

SNOW

	Wt. Lbs.
Fresh snow, 1 cu. ft. equals	5-12
Wet and compacted, 1 cu. ft. equals	15-50

ICE

	Wt. Lbs.
Ice, 1 cu. ft. equals	57½

APPROXIMATE WEIGHTS OF VARIOUS FARM PRODUCTS

	Lbs. per bu.
Barley	48
Beans	60
Blue Grass Seed	14
Bran	20
Buckwheat	50
Clover Seed	60
Corn, shelled	56
Corn, on cob husked	70
Corn, kaffir	56
Corn, unhusked	74
Cornmeal	50
Cotton Seed	32
Flaxseed	56
Hemp Seed	44
Millet	50
Oats	32
Onions	57
Peanuts	22
Peas	60
Popcorn	56
Potatoes	60
Potatoes, sweet	55
Rye	56
Timothy Seed	45
Tomatoes	56

FARM PRODUCTS (Continued)

	Lbs. per bu.
Turnips	55
Wheat	60

WEIGHTS PER CUBIC FOOT

Coal	50 lbs.
Barley	39 lbs.
Corn Shelled	45 lbs.
Corn on Cob	30 lbs.
Cotton	93 lbs.
Flax	93 lbs.
Hay	4 lbs.
Hemp	93 lbs.
Oats	26 lbs.
Potatoes, Piled	42 lbs.
Rye	45 lbs.
Wheat	48 lbs.

WEIGHTS OF ANIMALS

Calf	150 lbs.
Cow	1000 lbs.
Hog	240 lbs.
Horse	1300 lbs.
Lamb	75 lbs.
Sheep	110 lbs.
Steer	1200 lbs.
Man	150 lbs.

SIZES AND WEIGHTS OF PACKAGES

Eggs	30 x 12 x 12"	30 dozen	52 lbs.
Milk	14" dia. 26" high	10 gallon	115 lbs.
	18¼ x 14¼ x 12"	12 qt. case	63 lbs.
		Empty with bottles	33 lbs.
	18¼ x 14¼ x 10"	20 pt. case	54 lbs.
		Empty with bottles	33 lbs.
Butter	13½ x 12½"	Tub	30 lbs.
	16½ x 15"	Tub	60 lbs.
Cheese	16⅛ x 6⅜"	Box	30 lbs.
	15½ x 15"	Box	60 lbs.
Hay	46 x 30 x 26"	Standard Bale	210 lbs.
	43 x 24 x 18"	Small Bale	120 lbs.
Straw	43 x 30 x 26"	Standard Bale	180 lbs.
Cotton	27 x 27 x 54	Standard Bale	515 lbs.

SIZES AND WEIGHTS OF BARRELS

Apples	20⅜" Bilge	17⅛" head	28½" stave	150 lbs.
Flour	21 " Bilge	18 " head	28½" stave	200 lbs.
Sugar	25 " Bilge	20½" head	30 " stave	300 lbs.
Salt	21 " Bilge	18 " head	29 " stave	280 lbs.

RULE FOR CALCULATING WEIGHT OF STEEL BARS, ETC.

Having the number of cubic inches in any bar, to find its weight in steel, multiply by the decimal .2833 if exact weight is desired; or multiply by .3 if a practical weight, such as you are likely to get, is wanted. For the weight of wrought iron, multiply the cubic inches by .28. For the weight of cast iron, multiply the cubic inches by .26.

WEIGHT OF BAR STEEL PER FOOT

Size in inches.

SQUARE		ROUND		OCTAGON	
Size	Lbs.	Size	Lbs.	Size	Lbs.
1/8	.05	1/8	.04	1/8	.04
3/16	.12	3/16	.09	3/16	.10
1/4	.21	1/4	.17	1/4	.18
5/16	.33	5/16	.26	5/16	.28
3/8	.48	3/8	.38	3/8	.40
7/16	.65	7/16	.51	7/16	.54
1/2	.85	1/2	.67	1/2	.70
9/16	1.08	9/16	.85	9/16	.89
5/8	1.33	5/8	1.04	5/8	1.10
11/16	1.61	11/16	1.27	11/16	1.33
3/4	1.92	3/4	1.50	3/4	1.58
13/16	2.24	13/16	1.76	13/16	1.83
7/8	2.60	7/8	2.04	7/8	2.16
15/16	3.06	15/16	2.35	15/16	2.48
1	3.40	1	2.67	1	2.82
1 1/8	4.30	1 1/8	3.38	1 1/8	3.56
1 1/4	5.31	1 1/4	4.17	1 1/4	4.40
1 3/8	6.43	1 3/8	5.05	1 3/8	5.32
1 1/2	7.65	1 1/2	6.01	1 1/2	6.34
1 5/8	8.98	1 5/8	7.05	1 5/8	7.32
1 3/4	10.40	1 3/4	8.18	1 3/4	8.64
1 7/8	11.90	1 7/8	9.38	1 7/8	9.92
2	13.60	2	10.71	2	11.28
2 1/8	15.40	2 1/8	12.05	2 1/8	12.71
2 1/4	17.20	2 1/4	13.60	2 1/4	14.24
2 3/8	19.20	2 3/8	15.10	2 3/8	15.88
2 1/2	21.20	2 1/2	16.68	2 1/2	17.65
2 5/8	23.50	2 5/8	18.39	2 5/8	19.45
2 3/4	25.70	2 3/4	20.18	2 3/4	21.28
2 7/8	28.20	2 7/8	22.06	2 7/8	23.28
3	30.60	3	24.10	3	25.36
3 1/8	33.18	3 1/8	26.12	3 1/8	27.50
3 1/4	35.90	3 1/4	28.30	3 1/4	30.28
3 3/8	38.64	3 3/8	30.45	3 3/8	32.10
3 1/2	41.60	3 1/2	32.70	3 1/2	34.56
3 5/8	44.57	3 5/8	35.20	3 5/8	37.05
3 3/4	47.80	3 3/4	37.54	3 3/4	39.68
4	54.40	4	42.72	4	45.12
4 1/4	61.40	4 1/4	48.30	4 1/4	50.84
4 1/2	68.90	4 1/2	54.60	4 1/2	56.96
4 3/4	76.70	4 3/4	60.30	4 3/4	63.52
5	85.00	5	66.80	5	70.60
5 1/4	93.70	5 1/4	72.60	5 1/4	77.80
5 1/2	102.80	5 1/2	80.80	5 1/2	85.15
5 3/4	112.40	5 3/4	88.30	5 3/4	93.12
6	122.40	6	96.10	6	101.05

METRIC CONVERSION TABLES

U. S. TO METRIC

Linear

1 inch = 25.4001 millimeters
1 foot = 0.304801 meters
1 yard = 0.914402 meters
1 mile = 1.60935 kilometers

Square

1 square inch = 6.452 square centimeters
1 square foot = 9.290 square decimeters
1 square yard = 0.836 square meters

Cubic

1 cubic inch = 16.387 cubic centimeters
1 cubic foot = 0.02832 cubic meters
1 cubic yard = 0.765 cubic meters

Weight

1 grain = 64.7989 milligrammes
1 av. ounce = 28.3495 grammes
1 troy ounce = 31.10348 grammes
1 av. pound = 0.45359 kilogrammes

Capacity

1 fluid drachm = 3.70 cubic centimeters
1 fluid ounce = 29.57 milliliters
1 quart = 0.94636 liters
1 gallon = 3.78544 liters

METRIC TO U. S.

Linear

1 meter = 39.3700 inches
1 meter = 3.28083 feet
1 meter = 1.09361 yards
1 kilometer = 0.62137 miles

Square

1 square centimeter = 0.1550 square inches
1 square meter = 10.7640 square feet
1 square meter = 1.196 square yards

Cubic

1 cubic centimeter = 0.0610 cubic inches
1 cubic meter = 35.314 cubic feet
1 cubic meter = 1.308 cubic yards

Weight

1 milligramme = 0.01543 grains
1 kilogramme = 15432.36 grains
1 hectogramme = 3.5274 av. ounces
1 kilogramme = 2.20462 av. pounds

Capacity

1 milliliter = 0.27 fluid drachms
1 centiliter = 0.338 fluid ounces
1 liter = 1.0567 quarts
1 dekaliter = 2.6417 gallons

WIRE GAUGES

Sizes in decimal parts of an inch

No. of Wire Gauge	American or Brown and Sharpe	Birmingham or Stubs' Iron Wire	Pounds per foot 1 inch wide B. W. G.	Washburn and Moen	Pounds per 100 feet round W. & M. G.	Polished Drill or Stubs' Iron Wire	No. of Wire Gauge
0000	.46	.454	1.512	.3938	40.94		0000
000	.40964	.425	1.415	.3625	34.73		000
00	.3648	.38	1.265	.3310	29.04		00
0	.32486	.34	1.132	.3065	27.66		0
1	.2893	.3	1.000	.2830	21.23	.227	1
2	.25763	.284	.946	.2625	18.34	.219	2
3	.22942	.259	.863	.2437	15.78	.212	3
4	.20431	.238	.793	.2253	13.39	.207	4
5	.18194	.22	.733	.2070	11.35	.204	5
6	.16202	.203	.676	.1920	9.73	.201	6
7	.14428	.18	.600	.1770	8.03	.199	7
8	.12849	.165	.550	.1620	6.96	.197	8
9	.11443	.148	.493	.1483	5.08	.194	9
10	.10189	.134	.446	.1350	4.83	.191	10
11	.090742	.12	.400	.1205	3.82	.188	11
12	.080808	.109	.363	.1055	2.92	.185	12
13	.071961	.095	.316	.0915	2.24	.182	13
14	.064084	.083	.276	.0800	1.69	.180	14
15	.057068	.072	.240	.0720	1.37	.178	15
16	.05082	.065	.217	.0625	1.05	.175	16
17	.045257	.058	.193	.0540	.77	.172	17
18	.040303	.049	.165	.0475	.58	.168	18
19	.03589	.042	.140	.0410	.45	.164	19
20	.031961	.035	.117	.0348	.32	.161	20
21	.028462	.032	.107	.03175	.27	.157	21
22	.025347	.028	.093	.0286	.21	.155	22
23	.022571	.025	.083	.0258	.175	.153	23
24	.0201	.022	.073	.0230	.140	.151	24
25	.0179	.02	.067	.0204	.116	.148	25
26	.01594	.018	.060	.0181	.093	.146	26
27	.014195	.016	.053	.0173	.083	.143	27
28	.012641	.014	.047	.0162	.074	.139	28
29	.011257	.013	.044	.0150	.061	.134	29
30	.010025	.012	.040	.0140	.054	.127	30
31	.008928	.01	.0333	.0132	.050	.120	31
32	.00795	.009	.0300	.0128	.046	.115	32
33	.00708	.008	.0266	.0118	.037	.112	33
34	.006304	.007	.0233	.0104	.030	.110	34
35	.005614	.005	.0167	.0095	.025	.108	35
36	.005	.054	.0133	.0090	.021	.106	36
37	.004453	....	.....	......	......	.103	37
38	.003965	....	.....	......	......	.101	38
39	.003531	....	.....	......	......	.099	39
40	.003144	....	.....	......	......	.092	40

The United States Standard Gauge, unlike most other gauge tables, is a weight gauge based upon the weight per square foot in ounces. The thickness of iron and steel to produce the desired weight per square foot is calculated on the basis of the established weight of wrought iron and steel per cubic foot.

It is to be noted that steel is denser or heavier than wrought iron—approximately 2 per cent—and therefore must be rolled thinner than iron to equal the same weight per square foot.

United States Standard Gauge for Sheet Iron and Steel

Gauge Number	Thickness of Iron	Thickness of Steel	Weight per square foot in ounces Avoir.	Weight per square foot in pounds Avoir.
4	.2344	.2298	160	9.375
5	.2188	.2145	140	8.75
6	.2031	.1991	130	8.125
7	.1875	.1838	120	7.5
8	1719	.1685	110	6.875
9	.1563	.1532	100	6.25
10	.1406	.1379	90	5.625
11	.125	.1225	80	5.
12	1094	.1072	70	4.375
13	.0938	.0919	60	3.75
14	.0781	.0766	50	3.125
15	.0703	.0689	45	2.8125
16	.0625	.0613	40	2.5
17	.0563	.0551	36	2.25
18	.05	.0490	32	2.
19	.0438	.0429	28	1.75
20	.0375	.0368	24	1.5
21	.0344	.0337	22	1.375
22	.0313	.0306	20	1.25
23	.0281	.0276	18	1.125
24	.025	.0245	16	1.
25	.0219	.0214	14	.875
26	.0188	.0184	12	.75
27	.0172	.0168	11	.6875
28	.0156	.0153	10	.625
29	.0141	.0138	9	.5625
30	.0125	.0123	8	.5
31	.0109	.0107	7	.4375
32	.0102	.0100	6½	.40625

WEIGHTS OF LIQUIDS

Name of Liquid	Weight in Pounds per Cubic Foot.	Weight in Pounds per Gallon.	Number of Gallons in 3 Tons.	Number of Gallons in 5 Tons.
Acid, muriatic (40%)	75.0	10.02	588.82	981.35
Acid, nitric (91%)	94.0	12.56	477.72	796.20
Acid, sulphuric (87%)	115.6	15.50	387.09	645.15
Alcohol, ethyl (100%)	49.0	6.55	916.08	1,526.88
Alcohol, methyl (100%)	50.0	6.88	872.09	1,453.45
Alcohol, pure	49.6	6.63	904.05	1,506.78
Benzene	46.0	6.14	987.19	1,628.66
Chloroform	95.0	12.70	472.44	787.41
Ether	46.0	6.14	987.19	1,628.66
Ether, sulphuric	45.0	6.01	998.31	1,663.88
Fuel-oil	54.6	7.20	833.31	1,388.88
Gasolene	45.0	6.01	998.31	1,633.88
Ice	56.0	7.48	802.11	1,336.89
Kerosene	52.3	7.00	857.13	1,428.57
Linseed oil	58.7	7.84	765.30	1,275.51
Lubricants	57.0	7.62	787.38	1,312.33
Lubricating oil	57.2	7.65	784.29	1,307.18
Lye, soda (66%)	106.0	14.17	423.42	705.70
Mercury	848.7	113.46	52.88	88.13
Molasses	93.5	12.50	480.00	800.00
Muriatic acid (40%)	75.0	10.02	588.82	981.35
Naphtha	43.1	5.76	1,041.66	1,736.10
Naphtha, No. 2	44.1	5.80	1,118.67	1,697.75
Nitric Acid (91%)	94.0	12.56	477.72	796.20
Oil	52.0	6.95	865.23	1,442.07
Oil, mineral	57.0	7.62	787.38	1,312.33
Oil, vegetable	58.0	7.75	774.18	1,290.32
Olive oil	57.5	7.72	778.77	1,297.95
Petroleum	54.8	7.32	819.66	1,366.12
Petroleum, refined	50.3	6.72	892.83	1,488.09
Rape oil	57.5	7.72	778.77	1,297.95
Rigolene	39.0	5.21	1,151.63	1,919.35
Sea water	64.0	8.55	698.73	1,169.59
Snow, freshly fallen	8.0	1.06	a5,660.37	a9,433.95
Snow, wet (compact)	30.0	4.01	b1,496.25	b2,483.90
Sulphuric acid (87%)	115.6	15.50	387.09	645.15
Tar	62.4	8.34	719.40	1,199.04
Turpentine	54.0	7.21	832.17	1,386.96
Turpentine oil	54.3	7.25	827.58	1,379.31
Vinegar	62.4	8.34	719.40	1,199.04
Water, pure	62.4	8.34	719.40	1,199.04
Whale oil	57.6	7.70	779.28	1,298.70
Wine	62.3	8.32	721.14	1,201.92

a There are 750 cubic feet in 3 tons and 1,250 cubic feet in 5 tons. *b* There are 200 cubic feet in 3 tons and 333.30 cubic feet in 5 tons.

WEIGHTS OF ORES.

Name of Ore.	Weight in Pounds per Cubic Foot.	Weight in Pounds per Cubic Yard.	Number of Cubic Feet in 3 Tons.	Number of Cubic Feet in 5 Tons.
Blende	253	7,831	23.7	39.5
Brown hematite	237	6,399	25.2	42.1
Brown iron ore	237	6,399	25.2	42.1
Cassiterite	418	11,286	14.3	23.9
Chalcopyrite	262	7,074	22.8	38.1
Copper	262	7,074	22.8	38.1
False galena	253	7,831	23.7	39.5
Galena	465	12,555	12.9	21.5
Hematite	325	8,775	18.4	30.7
Iron (average)	281	7,587	21.3	35.5
Lead	465	12,555	12.9	21.5
Limonite	237	6,399	25.2	42.1
Magnetite	315	8,405	18.4	31.7
Manganese	259	6,993	20.8	34.7
Pyrites, copper	262	7,074	22.8	38.1
Pyrolusite	259	6,993	20.8	34.7
Red hematite	325	8,775	18.4	30.7
Sphalerite	253	7,831	23.7	39.5
Tin	418	11,286	14.3	23.9
Zinc	253	7,831	23.7	39.5

WEIGHTS OF MISCELLANEOUS MATERIALS.

Name of Material.	Weight in Pounds per Cubic Foot.	Weight in Pounds per Cubic Yard.	No. of Cubic Yds. in 3 Tons.	No. of Cubic Yds. in 5 Tons.
Beeswax	60.5	1,633	3.66	6.10
Cork	15.0	405	14.79	24.68
Fats	58.0	1,566	3.82	6.38
Flour, loose	28.0	756	7.93	13.22
Flour, pressed	47.0	1,269	4.72	7.88
Garbage	42.6	1,150	5.21	8.68
Hay	20.0	540	11.11	18.51
Lard	59.0	1,593	3.76	6.25
Leather	59.0	1,593	3.76	6.25
Paper	58.0	1,566	3.82	6.38
Powder, shaken	62.3	1,682	3.56	5.93
Rosin	68.6	1,852	3.23	5.39
Rubber	59.0	1,593	3.76	6.25
Rubber goods	94.0	2,538	2.36	3.93
Rubbish (rags, etc.)	7.4	200	30.00	50.00
Straw	20.0	540	11.11	18 51
Street sweepings	31.5	850	6.99	11.76
Sulphur	125.0	3,375	1.77	2 96
Wool, pressed	82.0	2,214	2.71	4 51

ANGLES OF SLIDE OF VARIOUS MATERIALS.

Name of Material.	Angle of Slide in Degrees.	Name of Material.	Angle of Slide in Degrees.
Ashes, dry	33	Earth, loose	28
Ashes, moist	36	Earth, packed	50
Ashes, wet	30	Garbage	30
Asphalt	45	Gravel	40
Brick	40	Ore	30
Cinders, dry	33	Ore, run-of-the-mine	35–40
Cinders, moist	34	Rubble	45
Cinders, wet	31	Sand, dry	40
Cinders and clay	30	Sand, No. 2, & broken stone	27
Clay	45	Shingles	40
Coal, anthracite	25	Stone	30
Coal, bituminous	30	Stone, broken	27
Coke	23	Stone, crushed	30
Concrete	30	Stone, run-of-the-mine	35–40

DEVELOPMENT OF THE MODERATE-PRICED CLOSED CAR

The far-sighted body engineer realized a long time ago that the ultimate development of the automobile would be a completely enclosed car. This car would have a permanent roof and glass side windows, either adjustable for ventilation, or removable for fair weather traveling. Development work, along this line, for providing all-season bodies has been going on for years, and a large number of body types have been made and manufactured.

There have been a good many factors which have governed the engineer in the past, and there are still a lot of conditions to consider. Some of the factors are engineering problems, some are selling and some might be termed psychological. The engineer was not, as a rule, expected to worry about selling problems or psychology, for the management, which included the sales division, considered this their own field. This condition is very well remembered by automobile engineers who have worked through the development of the automobile from its first conception. Psychology and sales influence did not affect the car in its early struggle, because the problem was to make the car run and keep it running. When this had been accomplished, the management and sales managers began to make the car suit their plans.

About this time the exterior appearance became an important factor, and inasmuch as the body engineer controls the general appearance of the car he had to work close to the management and produce what they wanted. "Pleasure" cars were being built and the industry was in its long clothes, and still using the nursing bottle. Manufacturers were so busy filling the demand for cars that little heed was paid to tomorrow until the war conditions and the government decided that the automobile was a luxury, and acted accordingly. In à very short time the industry grew up and found that it had the name of producing "luxuries," whereas it had made a very essential mechanical contribution to the world's advancement. By changing the word from "pleasure" to "passenger" car the useful side was emphasized, but to the public who were using cars it was evident a long time that the car was not a luxury in the true sense of the word, any more than a house, a train or trolley was.

The industry was developed on the open car, with the closed car having the status of a city, or limited use and "luxury" vehicle. Scientific and large production manufacturing brought the price of the cars down and rendered it possible to sell a closed car for a price which made it feasible for a large percentage of car users to obtain them. In the early stages the closed car was practically a hand-made and so-called "custom shop" product. While a large percentage of people stored their cars in the winter months, an open car satisfied their requirements, but the perfection of the mechanical construction made it possible to use the car the entire year around; so better protection and more comfort were demanded for winter driving.

Closed car development is only in its infancy. Originally the closed car was used only in the cities and on good roads at a low speed. The car of today must be capable of withstanding high speed driving on all kinds of roads. This means it must be substantially built to stand terrific shocks which a car will get in all-round driving. The public is car-wise, and would buy the closed car if they can purchase adequate transportation for a fair cost. The best value for the dollar will always get the market, but dollar value must also be consistent with general appearance, so the car must have a pleasing and attractive exterior, comfortable interior, and be constructed to cover the car's requirements, which have been proven acceptable in the development of the automobile. In other words, the doors should allow the passengers to get in and out easily; plenty of leg room and head clearance; comfortable seating arrangement; good vision and proper ventilating facilities are necessary.

The situation boils itself down to whether

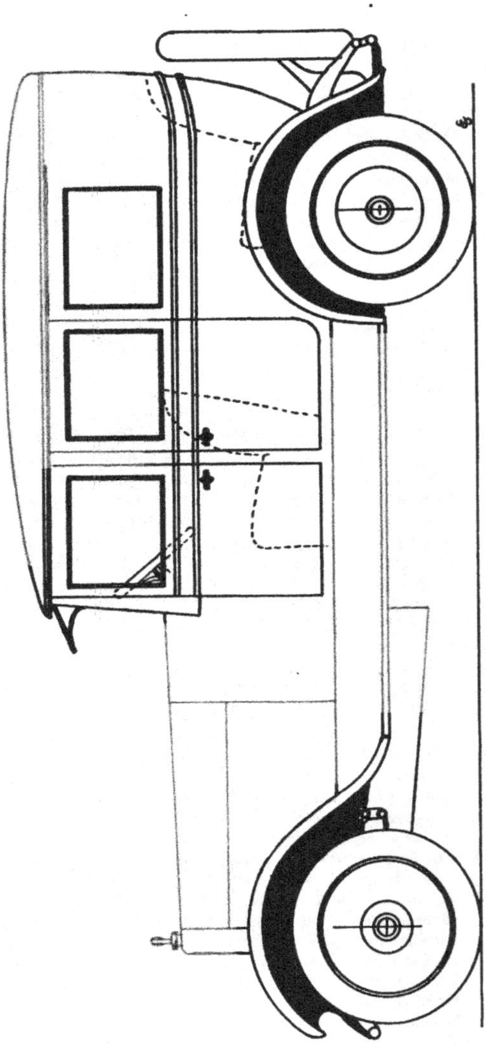

FIG. 1. Closed car laid out to show how the glass sizes may be cut down without affecting the car's utility. The roof has a deep, rounded side, which cuts off about two inches of glass all around and does not affect head room or vision. The belt line also has been raised, and this saves two inches of glass on the bottom.

the present type of closed car can be made cheap enough and in quantities large enough to replace the open car entirely, or will a new type closed car body be developed.

Take any popular-priced six-cylinder car and examine the five-passenger closed body and the five-passenger open body. There is a difference of about $700 between them. This difference is practically all in the body, because the chassis is usually the same, with the exception of the rear springs. The following list will permit an examination of the main differences between an open and closed car:

	Open Car	Closed Car
Body wood construction, lumber required	130 b. ft.	350 b. ft.
Mall. iron braces	30 lbs.	60 lbs.
Sheet steel	20 lbs.	15% more
Glass	5.3 sq. ft.	33 sq. ft.
Hardware door locks	light door lock	heavy lock.
Outside Yale lock	none	Yale lock
Door handles	light	heavy
Door dovetails	light	heavy
Door bumpers	light	heavy
Strikes	light	heavy
Window regulators	none	All doors 1 each door
Door hinges	two on each door	three
Door light	none	yes
Dome switch	none	yes
Sunshade fittings	none	yes
Fittings to operate rear quarter light	none	yes
Upholstery cushion springs	light	special
Back springs	standard	standard
Material for backs and cushions	leather	fabric
Door covering	imitation leather	fabric
Head lining	none	fabric
Side trimming	leather	fabric
Carpet, front	none	yes
Carpet, rear	medium grade	good grade
Foot rest	good	first-class
Robe rail	metal	fabric, metal fittings
Door straps	leather or cotton	fabric and cotton
Pull-to handle	none	metal or rope type
Hand pad	leather	none
Tops	cloth and metal sockets	Part of construction wood frame and compo. board
Drip moulding	none	aluminum
Painting	More time and coats on closed body, also more surface, requiring more paint	

This list is interesting for a general comparison between the open and closed bodies. and the first question is, what can be dispensed with without sacrificing utility. One thing that must be kept in mind is that labor. which will be a considerable item, does not show in this list, neither does the overhead, which includes factory and selling expenses, also dealers' profits. Every item which is simplified or omitted will reduce the overhead charge in proportion to the extent of the changes made. By examining the items separately some idea can be obtained for reducing cost.

1. Wood construction depends upon body lines. By simplifying the lines less lumber will be used, because the parts, especially the posts, can be cut out of small stock lumber. Costs increase considerably as the stock required becomes larger than 4 x 4. For the main construction it will not pay to use inferior or an unsuitable kind for the sake of economy.

2. Malleable iron. This material is used for braces. These could be closely analyzed and lightened where possible, but strength should not be sacrificed under any consideration.

3. The amount of steel for the body panel is entirely governed by the size of the body The weight should be kept to the minimum by using the gauge which experience has shown practical.

4. Glass. The amount of glass is governed by the general design. The design can be worked out to have a high belt rail and low top rail. This will cut down the length of the glass. The width can be controlled by making the body as compact as possible. The rear quarter glass can be made short. The rear back light can be considerably reduced. An open car can get along with about 130 square inches, whereas the closed car uses about 500 square inches.

In proportioning the glass size, economy can be effected by keeping in mind that glass is sold in even dimensions, increasing by 2 inches at a time. For instance, a glass measuring 21 x 19 will cost the same as one 22 x 20, and the size should be either increased to take the advantage of the amount which has to be purchased, or reduced to 18 x 20. There could be 80 square inches saved by reducing the size to 18 x 20.

5. Door locks can be simplified for closed body use.

6. Yale locks could be eliminated, as they are not used on open cars; switch locks and transmission locks would, no doubt, be used in any case.

7. The door handle could be simplified.

8. Door dovetails. This is a very important detail and should not be changed if its efficiency is impaired in any way.

9. Door bumpers. Small item—there is not much to be done with these.

10. Strikes can be made of malleable iron and lightened.

11. Window regulator. Straps could be used if felt desirable from sales standpoint, or a simple operating device similar to a Clement type regulator used in some cars on the rear quarter light.

12. Door hinges. The outside type forged steel can replace expensive concealed ones on the low-priced closed car.

13. Dome light. A convenience, not absolutely necessary; none are used on open cars.

14. Switch could be eliminated by hooking wire on to the instrument light.

15. Sunshade. Very useful and can be made very cheaply.

16. Rear quarter light could be made permanent. In most designs this light only drops a short distance and is very rarely used for ventilating. This light can be reduced in size.

17. Cushion springs. Same springs as are used in open cars could be used in closed ones.

18. Back spring. Same springs as are used in open cars could be used.

19. Trimming material. Cloth or leather can be used. Cost will be affected by market conditions.

20. Door covering. Cloth or imitation leather to suit back and cushion material.

21. Head lining, coated drill or cloth.

22. Side trimming. Imitation leather, real leather, or cloth to suit general trimming. The sides of the body and posts could be left untrimmed and painted instead. This is a doubtful economy, because the workmen would have to be carefully watched. The posts would have to be finished very carefully. Painting the posts and panels between would not be a very easy task, and when it is all done it is questionable whether there would be any economy over the trimmed style.

23. Use rubber mats or linoleum for economy.

24. Carpet should be used.

25. Foot rest design cheapened.

26. Robe rail. Rope type the most economical.

27. Door straps. Standard construction.

28. Door pull-to handle. Metal type can be used.

29. Hand pads. Applies to open cars only.

30. The tops of closed cars are covered with aluminum, composition materials and slats with imitation leather. Imitation leather on slats is the cheapest.

31. Drip moulding can be made of steel and screwed on. This is not the best method, but it can be used for economy.

32. Painting. The finish on the average car is none too good, but the closed body could have the same finish as the open cars; then there are the extra surfaces to be covered.

The system of manufacturing should also be carefully analyzed, and if a line system of progressive assembly is to be used, the design should be worked out to enable the best production results. One manufacturer makes various sections of the body and trims them, and then bolts the sections together. This is a new phase and should be closely watched. So far, the success of the closed body has been in its careful framing and jointing.

Trimming methods could be simplified without having sectional assemblies.

Detachable trimming could be made so it would only require a few screws, nails or tacks to assemble it in the car.

In Fig. 1 is shown a closed car laid out to show how the glass sizes could be cut down without affecting the car's utility. The roof, it will be noted, is like some of the old type cars and has a deep, rounded side. This cuts off about two inches of glass all around. It does not affect the head room or vision. The belt line has also been raised and this saves two inches of glass on the bottom.

HARMONY IN CAR UPHOLSTERY*

Harmony in car upholstery is a subject closely allied to color harmony in general; however, the decoration of car interiors is a new art in the sense that it is an old art newly applied.

A car interior requires different treatment from that accorded a house interior; in fact, we must approach the subject from an entirely different angle because of its more or less public nature. Furthermore, owing to quantity-production methods that have made possible the extensive use of the closed car, it was essential in the past to exert extreme caution in designing fabrics for car builders that would prove acceptable to the greatest number of different personalities. The result caused restraint in conception to be the guiding principle. Unfortunately this rule was not always recognized. The interior of a car has been treated at times as a piece of furniture to be upholstered as a boudoir with satins and hand-made lace, and all too often as a signboard whereon to write in blazing colors and patterns someone's ostentatious bad taste. It is time for a return to first principles. Good taste is an elusive quality, one of appraisal rather than of creation. The color scheme is the prime consideration in the correct decoration of motor-car interiors. The fabrics to be used are of importance also, but I will discuss the subject of color first.

In a broad sense color includes light and shade, and it is synonymous with light. It is safe to say that no subject has been more abused than color and I shall endeavor to clarify the entire subject of color harmony, or at least to convey a rudimentary knowledge of the subject so that the underlying principles, as applied to car interiors, can be understood.

The chief use of color is to beautify and, by its harmony, to appeal to the esthetic instinct. I believe that all persons possess esthetic taste, at least to a slight degree. We all know that taste is a matter of mental development. For instance, the child and the savage naturally prefer gaudy, brilliant colors, but more mature and refined persons prefer somewhat subdued shades and tints. Many persons possess a natural color-sense. However, color knowledge is attained easily in sufficient measure to enable one to determine readily which colors are harmonious, and which are not.

Color Harmony

Science maintains and proves every day that the orderly separation and co-ordination of simple facts will reduce the mysteries of yesterday to commonplace simplicities. Let us see what can be done toward simplifying the language of color and to define, in a measure, color harmony and its psychology.

We have the three pure colors of red, yellow and blue, as the basis of all other colors; they are elements in themselves, and cannot be produced by mixture. Therefore, they are called primary colors; each of them in its full intensity differs widely from the others in tone and quality. A number of other colors are in the spectrum and are termed binary or secondary colors. They are formed by mixing equal parts of two primary colors. For instance, when we mix equal parts of red and yellow, we produce a binary color, orange; yellow and blue mix to form the binary color, green; and blue and red mingle into a binary color, violet, thus completing the circle. We now have six colors and have widened the field considerably.

In the application of colors to decorative design, the binary or secondary colors are vastly more interesting than the primary colors. Any color with two component parts is more interesting than a purely elemental color. For instance, orange has greater decorative value than either yellow or red, green has more "quality" than either blue or yellow and violet is decidedly more interesting than either red or blue. Like the primary, the secondary colors can be combined effectively with any or all of the

* By R. S. Quaintance, Manager of Sales Promotion, Bridgeport Coach Lace Co., Bridgeport, Conn. A Paper read before the Body Section, Society of Automotive Engineers at the Annual Meeting in New York City, January 11, 1922.

neutral tones, white, black or gray. It is possible also at this stage to harmonize colors by using complementary or opposite colors.

Complementary colors are those which, if all colors were arranged around the circumference of a circle in proper order, would be exactly opposite the primary colors; in other words, by drawing a line from the red through the center of the circle, we would find that green is the complementary color. A primary and a complementary color form the strongest possible contrast to each other. It is not possible to find a color more different from red than green; they have nothing in common. However, these same colors have the peculiar power to enhance each other when placed side by side. Each seems to gain strength. A red apple looks redder when it nestles among the green leaves and the leaves look greener near the red apple. Another peculiar thing about the complementary colors orange and blue, yellow and violet and red and green, is the fact that they destroy each other when mixed together as in paints. A mixture of equal parts of any two complementary colors produces a neutral gray.

Color Psychology

To appreciate color harmony, it is necessary to know something of the psychology of color and the influence it exerts upon the emotions. It is a well-established fact that the entire personality is stimulated or depressed by colors, and very many data are available on the effects of visual color-stimulus upon the blood pressure, upon muscular, mental and nervous activity and upon the mood, and these effects are evident in many ways.

The esthetic significance of color was recognized many centuries ago; in fact, much of the psychology of color appears in mythology. Xenophon reports a conversation between Socrates and Parrhasius in which the esthetic value of color is shown to have been appreciated by the early Greeks. Plato also discusses the beauty and symbolism of color and reveals his acquaintance with its ability to excite emotional response. The influence of color is positive, and so is the difficulty to analyze it. Has not everyone felt the influence, the purity of freshly fallen snow and experienced a feeling of resentment or guilt when this beautiful white cloak is wantonly soiled or disturbed.

The meaning of colors, their language, already is established by common consent throughout the ages. We cannot ignore it. Not all can understand the causes of some of their sensations of happiness or comfort, but many of them are definitely traceable to the proper use of color. The recognition of beauty of color and the finer feeling toward colors are dependent in large measure upon the taste of the beholder. To repeat, taste is largely a matter of mental development. The eye loves color. Whether conscious of it or not, all people react or respond to the influence of color. Color has power to attract attention, to stimulate emotion, to animate and cheer or to quiet and depress.

Color is not as well understood as it should be for the reason that artists have maintained it as their own private preserve. They built a barbed-wire fence about it and put up a sign reading, "Thou shalt not trespass." They contend that, aside from a few general principles of color harmony, the realization of satisfactory color arrangements depends on esthetic instinct.

The phase of color study which the charts cannot touch is a kind of inner shrine that we may term its psychology. Wholly apart from color sources, dimensions or harmonies, what are the effects different colors exert on personal feelings and emotions? Why do we feel cheered and enlivened by light tones of color and depressed by the darker tones? Why, for instance, is red a more exciting color than blue? Why do we speak of colors as warm or cold when there is no physical sensation of heat or cold? Why are orange, yellow and red called advancing colors and blue and violet termed retreating colors? All these and many other questions are answered in the study of color psychology.

It has been known for centuries that colors are either warm or cold, and scientists have established definitely that some colors are retiring and others are advancing. Colors are either stimulating or depressing, as well as either warm or cold. It is seen readily, when looking at the spectral band, that colors on the red side are warm and the colors on the blue side cold. We think of sunshine as being warm. Yellow is the color of the sun; hence, a tint or shade containing yellow is called a warm tint or shade. It gives us the feeling of vitality and warmth. It is the symbol of action, of courage. Red stimulates and excites. When we are stirred with strong emotion, such as anger, bashfulness or hatred, the red blood leaps from our hearts and flames in our cheeks. Red is far more attractive than yellow and is conducive to the thought of heat, an element that is lacking in yellow. Who has not felt the cockles of his heart warm and expand before an open fire? In a room otherwise dark and gloomy, it seems a living thing. Gilbert K. Chesterton expresses our attitude toward an open fire in these characteristic words:

"A queer fancy seems to be current that an open fire exists to warm people! It exists to warm their hearts, to light their darkness, to raise their spirits, to cook their chestnuts, to tell stories to their children, to make checkered shadows on their walls and to be the red heart of a man's house and hearth, for which, as the great heathen said, 'a man should die'."

Color Values

Red is a warm color of great power. Yellow, orange, the browns and their neighboring hues and tints are warm colors. All of the brighter colors are symbols of light and warmth and in this sense yellow is gaudy, gay and enlivening and naturally emblematic of the sun.

Blue logically is associated with serenity, sedateness and cold calmness, and probably has been so associated through centuries of contemplation of the sky or heavens. Due to its being the color or the sky, it is characterized as dignified and soothing. It signifies melancholy as well as sedateness. Blue is everything that yellow and red are not. It is quiet, reserved and cold. We speak of the icy-blue stillness of the Far North, of cold, steely-blue eyes, and of having the "blues" when we are conscious of a lack of enthusiasm over life's affairs. In fact, coldness is the outstanding feature of blue and is communicated in varying degrees by all colors than contain blue components.

The cold colors are predominantly restful. However, they may be either agreeable or disagreeable. Their associations determine their agreeableness and these generally operate through sub-conscious channels. The warm colors generally are stimulating. But, as there are various degrees of stimulus, the warmer color also may be either agreeable or disagreeable.

Just why certain combinations of color are pleasing and others disagreeable, or even shocking, it is difficult to say. The question can be answered by stating that the combination is either harmonious or discordant. Harmony is merely a pleasing arrangement of colors; this is analogous to harmony in music, which is produced by pleasing arrangement of musical notes. In music, we know that rhythmic experiences or sounds are much more agreeable than those which are non-rhythmic. We must apply the same sort of rule to color. As an art, coloring is on an equal plane with music. However, as an exact science, music is far in advance.

Color Schemes

It is a well-established fact that the eye is pleased with a group of colors that show all of the primary colors in at least some degree. In color sensations the presence of all three colors appears to complete the color circuit. It is very difficult to explain this, and it must be accepted as a fact. As an instance, doubtless everyone has, at some time or other, tried the experiment of gazing intently at a round disc of strong red on a white background. If the red disc is suddenly removed and one continues to gaze at the white background, a green disc will appear. Green is a combination of yellow and blue. The eye has supplied the complement to the color that so filled it a moment before proving that certain colors seem to call for or demand certain other colors.

Color harmony cannot be achieved by selecting any two or more colors from the spectrum at random. It is essential that the colors chosen be at equal intervals from each other. We find, here, an analogy in music.

Two harmonious colors can be chosen from a color circle by selecting two complementary colors as has been described; that is, two colors exactly opposite to each other on the color circle are harmonious. I will now describe how to locate three-color harmonies by using an equilateral triangle to insure an equal interval between the color steps. If we place the triangle with the apex at yellow, the other two angles will locate the other colors in the color scheme, red and blue; and in traveling around the color circle, we locate the secondary and the neighboring colors. Having a proper regard for area, the three colors in each group can be used together in respect to the hue and luminosity of each color, in any value or intensity, and with gray, white or black, and can be depended upon to produce harmony. However, do not misunderstand me; the primary and secondary colors located by this simple method would not be suitable for use in closed-car interiors.

Another method of arriving at a color scheme that combines all the primary colors but in far more subtle proportions is by varying the interval. Just as there are different intervals in musical harmony, the majors and the minors, there are different intervals in color harmony. This gives us what is termed a split complement. The three points of this triangle can be made to travel around the color circle and locate the different split complements. These combinations contain also all of the elements of the straight complement or all of the primary colors; but they are present in different proportions.

Decoration of Car Interiors

It cannot be emphasized too strongly that, in the natural expression of refined taste, the purer color must be used sparingly and with great care. The tints and shades or tones are to be favored. The chief function of the purer color is emphasis only or, if one can so refer to it, the purer colors are for punctuations only. While it is agreed by most students that complementary colors are harmonious, there are certain conditions that must be considered, chief among which is area. For instance, green is the complementary of red. For certain purposes this combination is an excellent one. When applying it to a closed-car interior, we must neutralize the dominant hue.

Professor Munsell has compiled a color chart which shows that red is twice as strong as green in what he terms "chroma." In this instance we will regulate the area; so that our dominant color is green and considerably lowers the value of the red by darkening it. The green itself might well be grayed, preferably to an olive, and we then find, in applying the two complementary colors to a closed-car interior, that the color scheme has worked out to an olive-green cloth with a maroon stripe. It is a combination that is suggestive of spring and of the garb of nature and is representative of life, youth and freshness; a combination of warm and cold colors that is agreeable and restful, since the maroon stripe supplies sufficient life to tone-up the combination.

In passing it may be well to remark that nature supplies a wealth of harmony which must be studied to be appreciated. The student will observe quickly that nature employs a relatively small amount of pure colors. Even the beautiful sunsets are devoid of pure colors; all of the beautiful effects arise from ever-changing combinations of tints and shades.

A point to be borne in mind in regard to harmony in car upholstering is the fact that the less obvious the color element is, the more quality the different tones possess. This is applicable equally to color harmony in other fields.

The warmer tones are to be preferred for use in a closed car as against the colder tones. By those who have understood the subject, the warmer tints have been selected with a view to making the interior cheerful and inviting in its appeal, and to make it as pleasant as a well-appointed drawing-room in its season of greatest usefulness, winter. The same interior can be made cool, quiet and restful during summer, by the use of slip-covers having a color scheme based upon the cold side of the color circle. Slip-covers serve other purposes, chief among which is the protection of the upholstery from dust and dirt during periods in which the windows frequently are lowered when driving.

A still different kind of color harmony that I am sorry to say, is extremely popular with many car-builders, is what is termed a mono-chromatic group. This is made up of two or more tones of the same color. It is the most unobtrusive and conservative harmony scheme possible. It is always safe, but seldom interesting. There is, however, an even more severe treatment, that of employing a single solid tone. This practice requires painstaking care in the selection of laces, curtains, curtain cords and other trimming accessories to insure perfect matching I have seen some excellent examples of this type in the products of custom body-builders, but I do not commend its use to large manufacturers. This treatment would not wear well with some personalities; it would become extremely monotonous and probably tiresome.

In the mono-chromatic harmonies, those employing two or more tones of the same color and any other schemes except the solid tone, it is possible to give the interior that natural balance of light over darkness that is considered of prime importance in all decorative fields. In this case the floor covering would be of the darkest shade, the body cloth a trifle lighter, and the trimmings and head lining, if one is used, of a still lighter shade. In interiors of this nature it is possible to make the color scheme interesting by the use of one or more bright spots such as an enameled handle or vanity case, or by using other interior fixtures. This should be of a color complementary to the dominant color. If we must have mono-chromatic car-interiors, let us at least liven them up. If we insist that the cloth manufacturer supply us with fabrics containing two or more neutral shades, then let us *save* the interior; let us liven it up by the intelligent use of spots of color in the other interior fittings.

I wish to urge, however, that we be reasonable. Judging from the descriptions I have read of the cars shown at the recent Paris Salon, our European friends have achieved some remarkably grotesque discords in this line; apparently, this is a reaction from the recent war. Some of the cars shown must have been extremely ludicrous; nevertheless, we can learn much from the Europeans. Not all of their creations are to be classed as freakish; in fact, all Latins are more artistic than Saxons, as a rule and, generally, they are much more fond of color.

Fabrics

There are three general types of fabric, cotton, wool and mohair. Of the first, cotton, little need be said. Cottons are used generally in the cheaper cars only, although the cotton velours had an extensive run in the medium-grade cars during the war period, as did the cheaper cottons also. These velours, while giving a rich luxurious effect, are not really serviceable. The mohairs woven from the hair of the angora goat are extremely serviceable and have enjoyed several years of popularity. But the fabric that seems to be the most desirable from all angles of style, wearing quality and appropriateness is the various kinds of woolen cloths. They vary in weaves and weights and a particular type can be found for any upholstery purpose.

A point to be borne in mind when selecting fabrics is that any pronounced figure will soon grow extremely tiresome. The eye requires complete rest in a closed car, rest from the continuous motion outside. The influence of the interior should be one that is felt rather than seen; for this reason the brocades, tapestries and chintzes, or other furniture upholsteries, have no place in the correctly appointed motor-car. The pattern should be small and unobtrusive and appear in the body lining only.

Some manufacturers have experimented extensively with woolen cloths and today are weaving a cloth on looms especially designed for automobile fabrics. Special finishing machinery has been built to impart a broadcloth finish to all woolen cloths. A fabric that promises to become extremely popular is the new worsted cloth. Some of this is all worsted, and some grades are only worsted-faced. These cloths trim well and have unusual wearing qualities. Another new fabric that has great promise is the mohair sateen, a flat woven mohair. This can be made in a wide variety of shades and various patterns and, strange as it may sound, it is guaranteed by the manufacturer to outwear the car.

In conclusion, I want to enter a plea for more color; more color in interior and more color in the exterior. We literally have worked the funereal color, black, to death. For years we have painted the majority of our cars black. We have had too much of it. We should brighten the cars with color. As a result of the continued practice of using black, we have killed many an owner's pride in his car. When one drives home in a new car in these days, it is impossible to feel that one is creating any stir in the neighborhood. In fact, one's neighbor is apt to remark over the back fence, "I see you got the old bus washed up."

Recently, while in Havana, I was very agreeably surprised to note the difference of a long line of automobiles from that noticeable here. Every car in the line was clean and shiny and fully half of the cars were painted in colors other than black. They are very fond of color in Cuba but, to be frank, they know how to use it. A parade of automobiles in Havana is an interesting thing to watch.

Much attention has been devoted in the automotive industry to the subjects of line and form. Color has not yet played a leading part, but, sooner or later, color will come into its own and be on an equal plane with line and form. When this occurs and the automotive industry comes to an appreciation of the commercial value of beauty in color, we will find that color will not only supply the atmosphere and the drapery, but play a dominant role.

INDEX

Spotted something that isn't quite right?

Let us know and we'll fix it for future reprints.
And as a thank you for your diligence, we'll
send you a free Sothis book of your choice.

publisher@sothis.press

This page intentionally left blank

Made in United States
Orlando, FL
05 November 2026

90229010R00198